The Fundamentals of Signal Transmission

In line, waveguide, fibre and free space

Lem Ibbotson
former Lecturer, The Open University

ARNOLD

A member of the Hodder Headline Group
LONDON • SYDNEY • AUCKLAND

First published in Great Britain in 1999 by
Arnold, a member of the Hodder Headline Group,
338 Euston Road, London NW1 3BH

http://www.arnoldpublishers.com

British Library Cataloguing in Publication data
A catalogue record for this book is available from the British Library

ISBN 0 340 70576 0 (pb)

1 2 3 4 5 6 7 8 9 10

Commissioning Editor: Sian Jones
Production Editor: Wendy Rooke
Production Controller: Priya Gohil
Cover Design: Terry Griffiths

Typeset in $10\frac{1}{2}/13\frac{1}{2}$ pt Times by Academic & Technical, Bristol
Printed and bound in Great Britain by J. W. Arrowsmith Ltd., Bristol

What do you think about this book? Or any other Arnold title?
Please send your comments to feedback.arnold@hodder.co.uk

Contents

Foreword

The purpose of this book is to explain how signals are transmitted over significant distances using waves. A significant distance in this context is long compared to the wavelengths involved.

The subject is inherently mathematical, but a great deal of qualitative explanation is included to complement the mathematics.

The level of knowledge assumed is that acquired in the first year of undergraduate study in electrical engineering or applied physics, so the reader is expected to be familiar with electric circuit theory and to some extent with electric and magnetic field theory, and should be able to follow an argument involving differential calculus. Decibels are used throughout, but a discussion of this representation is included in the preface.

Chapters 1 to 6 give a general understanding of the field; chapter 7 gives further insights using a powerful graphical technique and finally there are a number of appendices included for those who want to go a little deeper into the mathematics.

Where relevant there are calculations posed at the ends of chapters which, if done, will help in understanding the material. Solutions are given at the end of the book.

Preface – Decibels

Communications engineers use decibels all the time. They were invented to represent power ratios, although they have since been used to express voltage ratios and a number of other ratios. They can also be used to represent absolute values related to an agreed base value.

As a power ratio, the decibel is defined by the relationship

$$X \text{ (dB)} = 10 \log_{10} (P_1/P_2)$$

and its inverse

$$P_1/P_2 = 10^{X/10}$$

The advantage of a logarithmic scale is twofold: first, the mathematical operation of multiplying is transformed into adding, and second, wide variations of value are compressed into a relatively small range of numbers.

Since the log to base 10 of 2 is 0.3010, a power ratio of two is approximately 3 dB. The logarithms of numbers less than 1 are negative and you will find if you try it on your calculator that $\log \frac{1}{2}$ is -0.3010, illustrating the fact that a minus number of dBs represents the inverse ratio of powers compared to the same positive number of dBs.

People who use decibels a lot tend to become adept at translating in their heads, so in power terms three dB is (approximately) twice as a ratio, six dB is four times, ten dB is ten times, twenty dB is a hundred times, fifty dB is a hundred thousand times, that is, ten to the fifth, and so on. What is fifty thousand times? Since it is half of a hundred thousand times, it is three dB less than fifty dB; in other words, forty-seven dB.

Rounding, on a logarithmic scale, has the same effect percentage-wise at all levels, thus, if you round from 46.8 to 47 dB you introduce exactly the same percentage error as if you round from 2.8 to 3 dB. In general, rounding to the nearest decibel gives about ten per cent accuracy, while rounding to the nearest tenth of a decibel gives about one per cent accuracy.

If two different voltages are applied to resistances of the same value, then the ratio of the powers delivered is V_1^2/V_2^2, so

$$X \text{ (dB)} = 10 \log(V_1/V_2)^2 = 20 \log V_1/V_2$$

This relationship tends to be used by electronic engineers even when the resistances are not the same, which can lead to errors if one is not very careful;

fortunately, however, since most transmission systems are matched to a constant impedance level, it generally gives no problems in the telecommunications context.

Absolute power levels can be defined by taking one of the powers in the ratio as a standard value. Suppose P_1 is the power to be defined and P_2 is taken as 1 watt: the value of the expression $10 \log P_1/1\,\mathrm{W}$ is given the designation dBW. For example:

7 dBW would be $10^{0.7} \approx 5\,\mathrm{W}$
-7 dBW would be $10^{-0.7} \approx 0.2\,\mathrm{W}$, which is $\frac{1}{5}\mathrm{W}$.

By taking the standard value as 1 milliwatt we can have 'dBm':

5 W in dBm would be $10 \log 5/0.001 \approx 37\,\mathrm{dBm}$

You may spot that this is 7 dB plus 30 dB to account for the factor of 10^3.

-37 dBm would be $10^{-3.7} \approx 0.0002\,\mathrm{mW}$,

which is $2 \times 10^{-7}\,\mathrm{W}$ or $\frac{1}{5} \times 10^{-6}\,\mathrm{W}$

In the last chapter, a further strange decibel 'flavour' appears – dBK. It turns out that it is necessary to take a gain, which is a ratio of powers, and divide this by a temperature in degrees Kelvin. The 'K' is thus simply an attempt to give information about what is represented; it is not to be taken as a unit with dimensional implications. If one requires to represent correctly the dimensions of the gain/temperature ratio, it is a power divided by a power and by a temperature, so its dimensions are K^{-1}. However, dBK^{-1} does not convey correctly the dimensions of the logarithmic ratio since it is not a decibel value divided by a temperature, so here it is always written dBK, although one will see both dBK^{-1} and dB/K in the literature.

Also in the last chapter you will find dBHz. The ratio so designated is that of a signal power to the noise power occuring in unit bandwidth. Again the 'Hz' is purely descriptive.

In the literature you will probably meet other decibel 'flavours'. The meaning is usually clear from the context.

1 Introduction: signals and waves

Before starting to explain the detailed processes by which signals can be transmitted, we need to be clear about exactly what a signal is. In essence, a signal is an encoded body of data that can be interpreted by an intelligent entity (person or computer). Let us see what is meant by *data* and by *encoded*.

1.1 DATA

Data is a collection of values for each of which there are just two alternatives; we can call it a set of 'yesses' and 'noes' or pluses and minuses or 'highs' and 'lows', but, as you are probably aware, it is usually represented by 1s and 0s. This representation is described as *binary data* because 1s and 0s in a number system where no other digits exist are called *binary digits* or *bits* for short.

Analogue data

The definition of data given above may seem odd, particularly if one is aware that much of the data received by the senses, and that captured by many older forms of sensing instruments, appears in analogue form, like, for instance, the continuously varying pressure of a sound on the eardrums. It will be shown in a later section that analogue data can be interpreted in terms of binary data.

Information

A body of data is transmitted in order to transfer one or more messages containing information. According to information theory, the amount of information in a message depends on the probability of that message being sent: the less likely the message the more information. In information theory the information in a message is also measured in bits and an ideal message would have the same number of bits of information as it has bits of data; practical messages have much less. Using information theory, methods have been developed to *compress* data so that less data has to be transmitted in a signal to transmit a given volume of information: further detail is outside the scope of this book.

Parallel data

Data can be transferred all at the same time – for example: 'Is there a voltage between earth and the red wire?', 'Yes'; 'Is there a voltage between earth and the blue wire?', 'Yes'; 'Is there a voltage between earth and the white wire?', 'No', and so on: this is described as *parallel data*. There is, however, a limit to how much data can be presented in parallel, so large volumes of data have to be presented sequentially; which brings us to the matter of encoding.

1.2   ENCODING

The simplest encoding is where binary values are presented in sequence: common sense suggests that sequential values should be presented for equal intervals of time. If the alternative values are the presence or absence of a voltage, then the voltage waveform (graph of voltage against time) of the data is as shown in Fig. 1.1. This is so simple that it is often described as *raw data*. The time duration, in this case, of a 1 or a 0 is known as the *symbol interval*, and the number of symbol intervals per second is called the *signalling rate* or *baud rate* since it is measured in units called *baud*. For raw data, the bit rate, that is the number of binary digits presented per second, is the same as the baud rate.

The raw data presentation as shown in Fig. 1.1 is known as *non-return-to-zero* (NRZ) to distinguish it from an alternative *return-to-zero* (RZ) mode shown in Fig. 1.2 for the same data sequence. Often in an RZ waveform the pulse duration of a '1' is half the symbol period.

Raw data is generally further encoded to make it suitable for transmission. This may involve inverting alternate 1 pulses and/or adding extra pulses. More radically, it frequently involves taking short sequences of pulses and replacing them by other symbols or sets of symbols. If, for example, a transmitted pulse is allowed to have four levels, as shown in Fig. 1.3, then each level could represent two binary digits as shown in the figure. Each of the three pulses, of different sizes, together with the absence of a pulse, is now a *symbol* and can be transmitted in one symbol interval. For a given data rate the signalling rate is halved, which

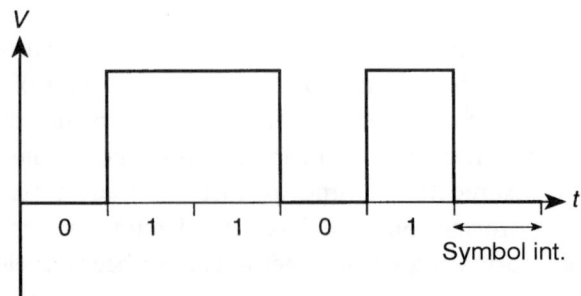

Fig. 1.1   An NRZ data waveform

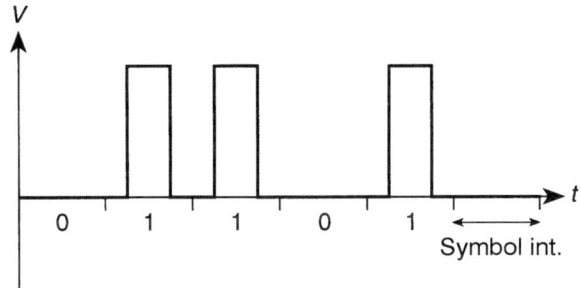

Fig. 1.2   An RZ data waveform

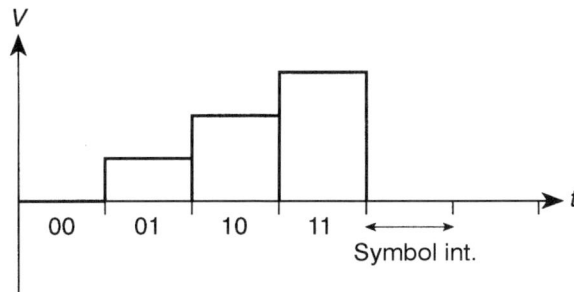

Fig. 1.3   A four symbol (quaternary) code

halves the bandwidth required (bandwidth will be dealt with later in this chapter), but the noise immunity is reduced. Finally, the data, in its raw form or encoded as above, may be used to modulate a sinusoid, that is to vary its amplitude, frequency or phase in step with the appropriate waveform. This modulation is yet a further step of encoding and results in symbol intervals containing symbols each of which is a section of sinusoidal voltage or electric field of a given amplitude, frequency or phase (or combination of these).

1.3   FOURIER'S THEOREM

Fourier's is a very general theorem which can be applied to any two connected physical quantities, but application of it here will be to the variation of an amplitude, typically a voltage or an electric field strength, with time. What the theorem tells us in this context is that the variation of the amplitude with time is equivalent to the sum of a number (usually an infinite number) of sinusoidal variations with time. This is very often illustrated using a square waveform as shown in Fig. 1.4. Figure. 1.4(a) shows a voltage varying sinusoidally with time. Figure. 1.4(b) shows another sinusoidal voltage of smaller amplitude and three times the frequency. Adding the two graphs gives a total variation of voltage with time shown in Fig. 1.4(c). Now add another voltage sinusoid of smaller amplitude still and five times the frequency of the first; shown in 1.4(d). The result, in Fig.

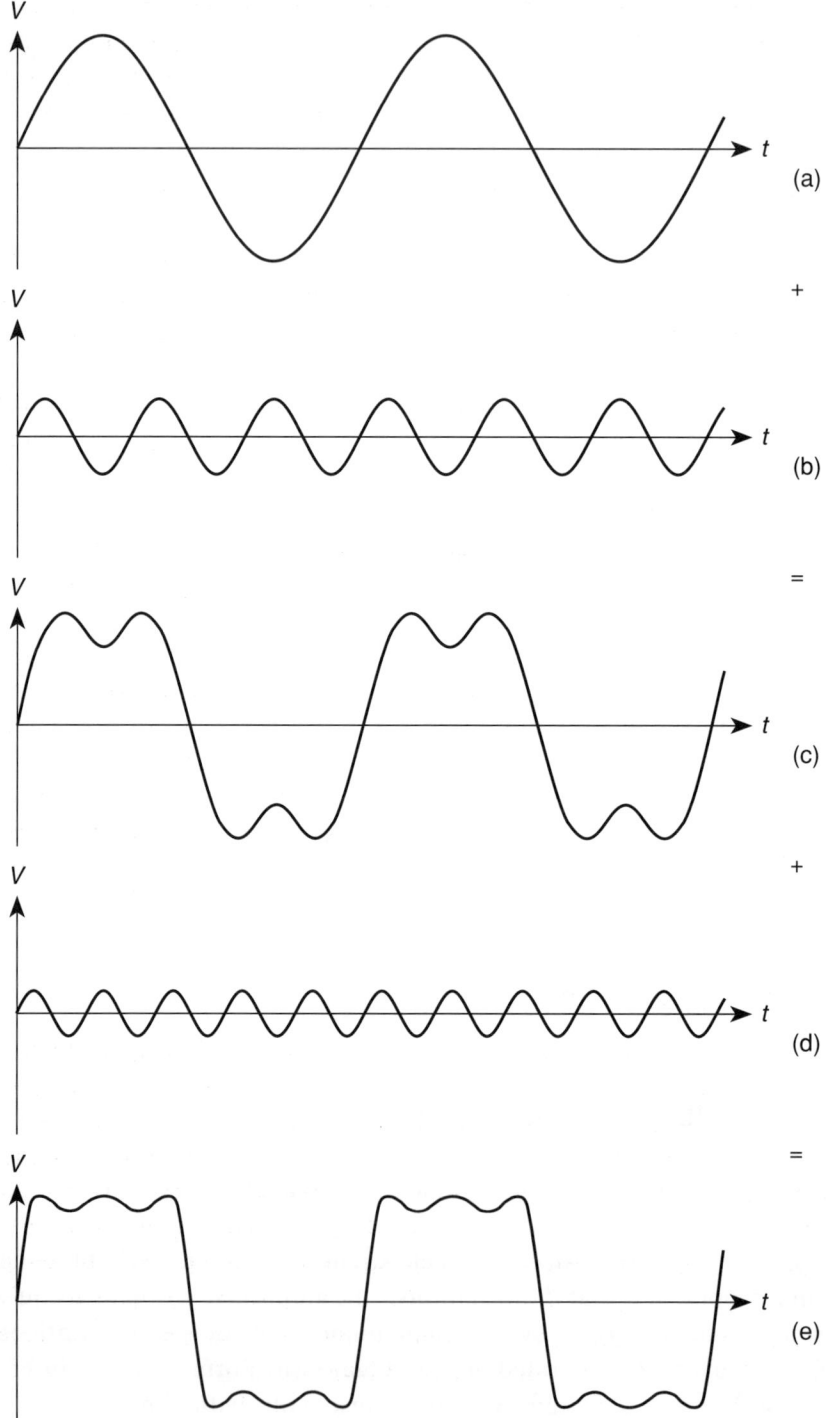

Fig. 1.4 The build-up of a square wave from sine waves

1.4(e), is beginning to look like a square waveform, and one will easily be persuaded that if this is continued, with components of higher and higher frequencies, the ultimate result is a square waveform. You may well guess that each component must be in the correct position – that is, must have an appropriate phase – and must have the correct amplitude. The lowest frequency sine waveform, which has the same frequency as the repetition rate of the square waveform, is called the *fundamental*. The sine waveform of three times this frequency is called a *third harmonic* and that at five times the fundamental frequency a *fifth harmonic*.

If an appropriate value of d.c. were added to the square waveform, setting the bottom of the waveform on the zero line, then this would be the waveform of raw data representing 1 0 1 0 1 0 . . .

With the addition of a d.c. component representing its average value, any repetitive waveform can be built up from a fundamental plus harmonics of appropriate phases and frequencies. Usually all harmonics are present; the even ones just happen to be absent in the case of a square waveform. Even if a varying amplitude does not repeat itself it can still be represented in this way: it can be thought of as having an infinite period – i.e. a repetition rate of zero – so that its fundamental frequency is zero and its harmonics are infinitely close together.

The set of sinusoids that together make up a signal is called its spectrum: a d.c. average value is thought of in this context as a zero-frequency component.

There are two reasons why this application of Fourier's theorem is important. The first is that we have very well-developed mathematical techniques for dealing with sinusoids. The second is that in all cases of interest we find that there are frequencies above which, and often also frequencies below which, we can in practice ignore the spectral components: this leads to the idea that any signal is equivalent to, or occupies, a (sinusoidal) frequency band.[1]

The sampling theorem

The concept of analogue data can now be reconciled with the original definition of data. The sampling theorem indicates that any signal of continuously varying amplitude, of which the spectrum has been restricted below a given highest frequency, can be represented by, and reconstructed from, a set of instantaneous values or *samples* taken at a rate no lower than twice that highest frequency. These samples can then be represented, to any desired degree of accuracy, by sets of binary digits (the greater the number of digits per sample the greater the accuracy). Hence the analogue signal can, in a sense, be thought of as a form of modulated digital data. In fact, nowadays many analogue signals are converted into digital form for further encoding and transmission.[2]

---

[1] The term *frequency* should strictly only be applied to sinusoids; for other repeating waveforms the term used should be repetition rate.

[2] It is essential that the spectrum of an analogue signal which is to be sampled be cut off at a frequency below half the sampling rate, otherwise spurious components appear in the reconstituted analogue signal by a process known as *aliasing*.

1.4 PRACTICAL SIGNALS AND THEIR SPECTRAL CONTENT

An analogue signal may have a d.c. content and will have an upper frequency limit above which the elimination of higher frequency components in the spectrum is not considered to produce significant degradation. Raw digital data also has a d.c. component and is usually considered to require the transmission of all frequencies up to a frequency equal to the signalling rate. Similarly, for encoded data other than that involving modulation of a sinusoidal carrier, the upper frequency transmission requirement is usually taken to be the signalling rate. Some signals may have a practical lower frequency limit as well, for example, high quality sound does not require the transmission of frequencies below about 70 Hz, and telephone speech is subjected to filtering which cuts out frequencies below 300 Hz. Some digital encoding schemes are designed to limit the low-frequency content of the signal spectrum.

In the case of the modulation of a sinusoidal carrier, any of the basic forms of modulation (amplitude, frequency or phase) results in the creation of *side bands*, that is two bands of frequency components, one above and the other below the carrier frequency, which are mirror images of each other with respect to the carrier. The components of a side band are related to the modulating signal in different ways, depending on the type of modulation, but in all cases components of significant amplitude are contained within a limited frequency range from the carrier. The frequency spectrum required to be transmitted may consist of either one or both of these side bands and may include the carrier or not. Where several sinusoidal carriers of different frequencies are modulated with the data of different messages (care being taken that their bands do not overlap) the spectrum of the total transmission consists of all these bands.

1.5 WAVES

In all practical systems, signals are transmitted by means of waves, so the next task is to explain what a wave is. A lot of confusion is caused by the fact that waves are often taught in elementary courses alongside simple harmonic motion, with the implication that these two topics are directly linked – they are not.

Time delay with distance

The essence of wave propagation is the transmission of a process from a point in a medium to the point next to it with a lapse of time. Think of a crowd at a football match: if a person raises their hands and then lowers them, and the person next to them does the same a little later, and the person next to them later again, the crowd generates a *Mexican wave*. There must be some mechanism which causes each point in the medium to follow the previous one – in the case of the Mexican wave it is seeing your neighbour carrying out the procedure.

Waveform

The propagating processes in the waves which will be of interest here are always changes in the amplitude of some property – changes, in fact, which represent the signal to be transmitted. In this context, the amplitude variations of the signal with time are called the *waveform*.

Sinusoidal waves

Since, as has been seen, a signal can always be represented as a group of sinusoids, it is sensible for the purposes of analysis to consider sinusoidal waves, that is, waves of which the property variation is a sinusoidal variation of amplitude. In this case, each point in the medium carrying the wave executes a sinusoidal cycle of amplitude with a phase which is delayed relative to the point before it. Thus, the first thing to understand about a sinusoidal wave is that there is a delay of phase in the direction of propagation of the wave, proportional to distance.

The driving point

A wave is set up in a medium by inserting power to induce and maintain the amplitude variation. This power is inserted, by the transmitter, at the *driving point*. Energy constantly moves away from the driving point along the direction of propagation of the wave.

1.6   SUMMARY

A signal consists of encoded data which, as presented to the input of a transmission system, can be regarded as the sum of a collection of sinusoids. The task of the transmission system is to transport these sinusoids to its output maintaining, as far as possible, their relative amplitudes and phases; this it does by propagating a wave with the signal as its waveform. Analysis of the wave propagation is carried out by considering separately the propagation of each of its sinusoidal components.

2 Twin-wire transmission line

As you read this chapter you may think that the detailed treatment given to this topic is more than its importance merits. The reason for the detail is that the simple analysis possible in the case of twin-wire line allows us to develop a number of concepts which will also be applicable to the other transmission media that are described in later chapters.

Twin-wire line is the oldest type of electrical signal transmission medium, but it is still very much in use in the form of twisted pair. For the moment assume a long straight pair of parallel wires – it will be explained later why they may be twisted.

The signal is applied by the transmitter as a voltage waveform between the wires at one end – the driving point. At the other end the receiver is connected; *see* Fig. 2.1. Assume that the receiver correctly terminates the transmission line – what this means will be discovered later.

The line is linear, in the electrical sense, meaning that a sinusoidal input gives rise to a sinusoidal output of the same frequency. Its electrical properties can be described in terms of series resistance and series inductance per unit length together with shunt conductance (sometimes called *leakance*) and shunt capacitance per unit length; rather as shown in Fig. 2.2, except that the components are not separate, but intermingled. The total resistance per metre of the two wires is designated R, the total inductance per metre L, the conductance per

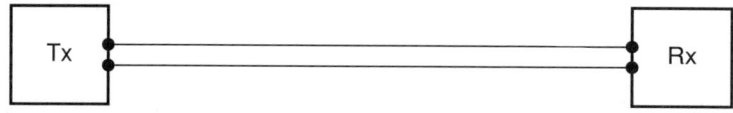

Fig. 2.1   Twin-wire line connecting a transmitter and a receiver

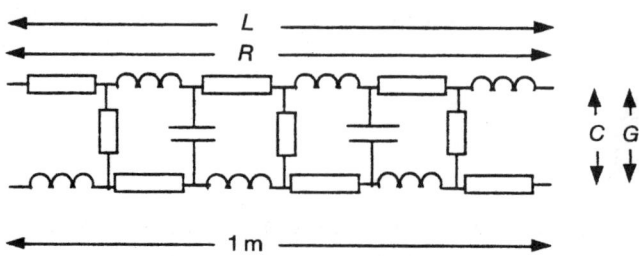

Fig. 2.2   Representation of a 1 metre length of line

metre between the wires G and the capacitance per metre C. These are known as the primary constants of the line (although R and G, in fact, vary with frequency).

For reasons explained in Chapter 1, we shall first analyse the electrical behaviour of the line with an input from a sinusoidal generator.

2.1   A TRAVELLING SINUSOIDAL WAVE ON A LOSS-FREE LINE

For the moment assume that the resistance and leakance of the line are both zero. Since capacitance and inductance do not absorb energy, a wave on such a line must travel without loss.

The input voltage can be described by the equation

$$v = V \sin(\omega t)$$

where ω is the angular frequency of the sinusoid (measured in radians per second) and V is its amplitude.

As the input voltage goes through its cycle it drives charge into and out of the beginning of the line cyclically, charging and discharging the capacitance between the two wires. At the same time the inductance of the line limits the current. The direct influence of the input voltage only extends a short way along the line, but the changing voltage on the first section in turn drives charge into and out of the next section of line and so on. A short time after the input has first been applied the line will settle to a *steady state* in which each point on the line, up to the receiver, has a sinusoidally varying voltage of the same amplitude and frequency as that of the input. However, the transfer of the signal from each part of the line to the next involves a time delay, so that at each point on the line the phase of its sinusoidal voltage is delayed relative to the previous point. Delaying a sinusoidal variation in time is equivalent to moving it backward in phase, so that the voltage can be described, at a distance x along the line from the generator, by the equation

$$v = V \sin(\omega t - \beta x) \tag{2.1}$$

β is called the *phase change coefficient* (its SI units are radians per metre).

A similar expression to Equation (2.1) holds for the current as a function of time and distance – as indicated in Fig. 2.3. The wave has both current and voltage associated with it; however, *the two are found to be in phase at all points and proportional in amplitude*, so the wave can conveniently be described in terms of the voltage alone.

Wave profile

If a graph is plotted of the voltage at different points over a section of the line at one instant of time (a voltage versus *distance* graph), the result is a sinusoidal shape; this graph is called the *wave profile*.

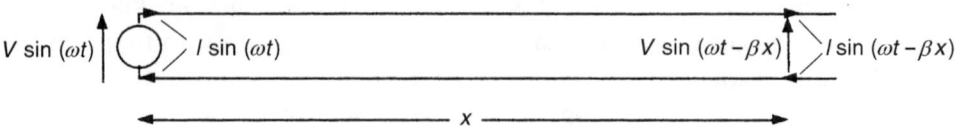

Fig. 2.3   Variation of sinusoidal current and voltage with distance on an ideal twin-wire line

How does the wave profile change with time? This is illustrated in Fig. 2.4 which shows the wave profile on the section of the line next to the generator at two instants – 2.4(a) when the generator voltage is going through zero, 2.4(b) about one-sixth of a cycle later.

In Fig. 2.4(a) the point P_1 on the line has just reached its peak instantaneous voltage. In 2.4(b) the voltage at P_1 has fallen, while P_2 has reached the peak. The whole waveform appears to have moved forward (that is, in the direction of increasing phase lag) without changing its shape.

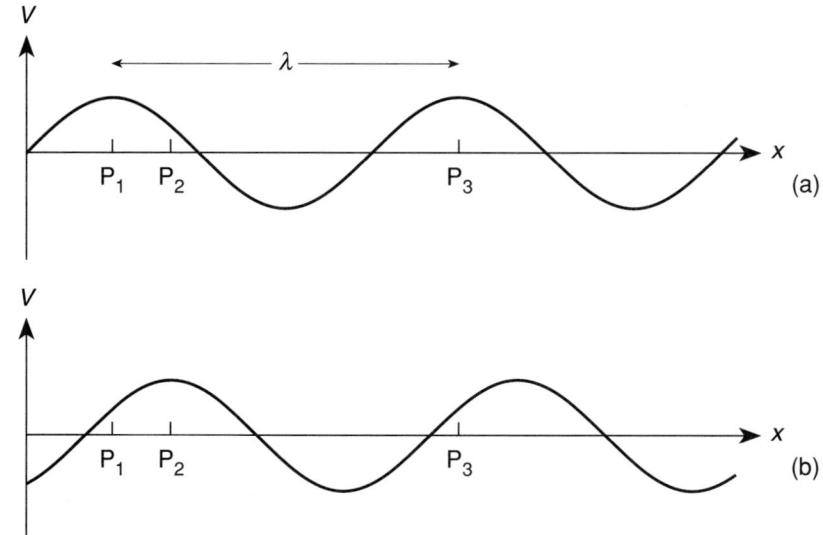

Fig. 2.4   Wave profile on a line at two times separated by 1/6 cycle

Phase velocity and wavelength

The wave profile, in fact, moves forward with time at a constant rate; this is called the *phase velocity* v_p.

The length of a complete cycle of variation on the wave profile is called a *wavelength*, symbol λ. In Fig. 2.4, points P_1 and P_3, for example, are λ apart. All points on the line separated by a distance λ have voltages in the same phase. Looking at the formula $v = V \sin(\omega t - \beta x)$, it follows that λ is the value of x which makes the phase delay 2π – so

$$\beta\lambda = 2\pi \qquad \text{i.e.} \qquad \beta = \frac{2\pi}{\lambda}$$

Consider the time it takes for the voltage at the point P_3 on the line, starting at the instant shown in Fig. 2.4(a), to go through one complete cycle. This length of time is one period, or $1/f$ where f is the generator frequency. During this time the wave profile moves a distance λ, so that the next peak comes to P_3. Hence

$$v_{\mathrm{p}} = \frac{\text{distance}}{\text{time}} = \frac{\lambda}{1/f} = f\lambda$$

Also, since $f = \omega/2\pi$ and $\lambda = 2\pi/\beta$

$$v_{\mathrm{p}} = \frac{\omega}{\beta}$$

Energy stored and transferred

A charged capacitor and an inductor carrying a current both have stored electrical energy. Hence it can be seen that a transmission line carrying a continuous wave has energy stored all along its length.

Energy is also *transported* by the wave. This can be understood by considering an infinite line fed by a sinusoidal generator switched on at some time in the past. The front of the wave disturbance proceeds along the line so that the wave extends into an extra length of line each second; the energy stored by this extra length of line must be provided by the generator. The speed with which energy is carried along the line by the wave is called the *group velocity*, v_{g}.

How does the group velocity compare with the phase velocity? It will be shown later that the group velocity is given by

$$v_{\mathrm{g}} = \frac{\mathrm{d}\omega}{\mathrm{d}\beta}$$

A line correctly terminated by a receiver is equivalent to a pipe for electrical energy: the generator drives energy in at a certain rate (the input power) and the receiver accepts energy at the same rate, while the line itself carries the energy on its way through.

2.2   A TRAVELLING SINUSOIDAL WAVE ON A PRACTICAL LINE

Attenuation

Energy is lost from a wave on a practical transmission line: in consequence the wave amplitude is attenuated with distance. The rate of loss of amplitude with distance is proportional to the amplitude, which means mathematically that the fall of amplitude is exponential, so instead of taking the amplitude of the voltage at any point x as the amplitude of the input voltage, V, it must be taken as $V\,e^{-\alpha x}$ where α is the *attenuation coefficient* (its SI units are nepers per metre). The equation relating voltage to distance and time on the line becomes

$$v = V e^{-\alpha x} \sin(\omega t - \beta x) \tag{2.2}$$

Again, an equation of similar form holds for current (but note, *on a practical line the current and voltage at the same point are not necessarily in phase*).

Since the attenuation of a line is usually quoted in decibels per unit distance, the relationship between the neper and the decibel should be indicated, as follows. The ratio of two voltage amplitudes one metre apart on a line, V_1 and V_2, can be written

$$V_1/V_2 = e^{\alpha}$$

V_2 being the voltage further along in the direction of propagation. If α is 1 neper per metre this becomes

$$V_1/V_2 = e$$

which in decibels is

$$20 \log_{10}(V_1/V_2) = 20 \log_{10} e = 8.686\,\text{dB}$$

One neper is 8.686 dB, so it is a rather large unit.

Phasor currents and voltages

Further analysis is facilitated by representing the currents and voltages on the line as phasors.

A phasor can be used to represent the amplitude of a sinusoidal voltage or current and its phase difference from a reference sinusoid of the same frequency (note that a phasor does not include any representation of the frequency). Thus, a voltage $V \sin(\omega t + \phi)$ has a phasor $\mathbf{V}$ which can be represented in amplitude-angle form as $V \angle \phi$; in component form as $a + jb$ where $a = V \cos \phi$ and $b = V \sin \phi$; or in complex-exponential form as $V\,e^{j\phi}$: it is this last form which is particularly useful here. (See Appendix 1.)

The phasor of the driving-point voltage is called $\mathbf{V}_0$. Referring to Equation (2.2) the phasor of voltage at distance x from the driving point can be written

$$\mathbf{V}_x = \mathbf{V}_0\,e^{-\alpha x} e^{-j\beta x}$$

which is the same as

$$\mathbf{V}_x = \mathbf{V}_0\,e^{-(\alpha + j\beta)x}$$

The expression $\alpha + j\beta$ is called the *propagation constant* and is given the symbol γ:

$$\gamma = \alpha + j\beta$$

so

$$V_x = V_0 e^{-\gamma x} \tag{2.3}$$

Similarly a current phasor at x can be related to the driving-point current:

$$I_x = I_0 e^{-\gamma x} \tag{2.4}$$

Remember, I_0 and V_0 are themselves phasors, and their angles are not necessarily the same.

The electrical properties of a short section of line, of length δx, at point x on the line will now be analysed. In Fig. 2.5 the resistance, the inductance, the leakance and the capacitance of this small section are lumped together so as to be able to draw the diagram, but remember that they are all distributed. The voltage changes from one end of the section to the other by δV and the current by δI.

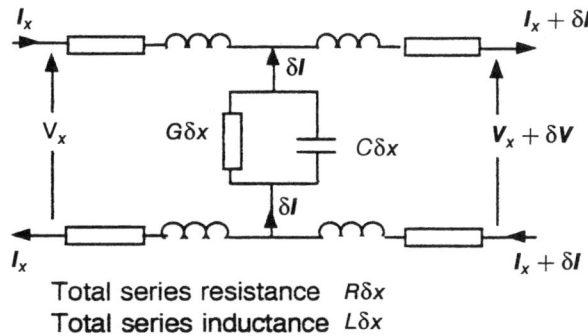

Total series resistance $R\delta x$
Total series inductance $L\delta x$

Fig. 2.5   Electrical properties of a length of line δx

Applying Kirchhoff's voltage law around the loop of the wires, and ignoring the small current variation

$$\delta V = -R\delta x I_x - j\omega L\delta x I_x$$

Dividing by δx and allowing $\delta x \to 0$

$$\frac{dV}{dx} = -(R + j\omega L)I_x \tag{2.5}$$

Similarly, considering the current between the wires and ignoring the small voltage variation

$$\delta I = -G\delta x V_x - j\omega C\delta x V_x$$

giving

$$\frac{dI}{dx} = -(G + j\omega C)V_x \tag{2.6}$$

Differentiating Equation (2.3)

$$\frac{dV_x}{dx} = -\gamma V_0 e^{-\gamma x} = -\gamma V_x$$

Combining this with Equation (2.5)

$$\gamma V_x = (R + j\omega L)I_x \tag{2.7}$$

Similarly from Equations (2.4) and (2.6)

$$\gamma I_x = (G + j\omega C)V_x \tag{2.8}$$

Characteristic impedance

Dividing Equation (2.7) by (2.8) gives

$$\frac{V_x}{I_x} = \frac{R + j\omega L}{G + j\omega C} \frac{I_x}{V_x}$$

which yields

$$\frac{V_x}{I_x} = \sqrt{\frac{R + j\omega L}{G + j\omega C}}$$

This equation shows that the ratio of voltage to current in the wave is the same at any point on the line: this ratio is called the *characteristic impedance* of the line – symbol Z_0 – so the equation can be rewritten

$$Z_0 = \sqrt{\frac{R + j\omega L}{G + j\omega C}} \tag{2.9}$$

For the ideal line which was discussed first, R and G are zero, so, from Equation (2.9)

$$Z_0 = \sqrt{\frac{L}{C}}$$

The fact that this term is real – i.e. Z_0 is resistive – indicates that, as stated earlier, the current and voltage of a travelling wave on an ideal line would be in phase.

More generally, the expression in Equation (2.9) is not real; Z_0 is partly reactive and the current and voltage are not in phase.

Theoretically there is a special case in which Z_0 is real even though R and G are not zero; it is if

$$\frac{R}{L} = \frac{G}{C}$$

so that Equation (2.9) can be written

$$Z_0 = \sqrt{\frac{L(R/L + j\omega)}{C(G/C + j\omega)}} = \sqrt{\frac{L}{C}}$$

This is known as the *distortionless condition*, but practical lines are never any-where near it – G/C is much smaller than R/L. In the past some lines carrying long distance signals were *lump loaded* with inductors at regular intervals to get the ratio R/L nearer to G/C, but this practice is now outdated.

L and C should be virtually frequency independent, but R and G are not simple d.c. resistance and conductance; the value of R is influenced by the *skin effect*, while the value of G is accounted for in part by dielectric hysteresis losses. These effects vary with frequency in such a way that R and G increase as ω increases, but less than proportionally. Hence there is a frequency level above which we can assume that

$$\omega L \gg R \qquad \text{and} \qquad \omega C \gg G$$

(The criterion $\gg$ (very much greater than) is generally taken, in this context, to mean an order of magnitude greater – i.e. at least 10 times.)

Under these circumstances, in Equation (2.9) take

$$R + j\omega L \approx j\omega L$$

$$G + j\omega C \approx j\omega C$$

so $Z_0 \approx \sqrt{L/C}$ as for the loss-free case.

This is the value of characteristic impedance that cable manufacturers normally quote.

Propagation constant

Multiplying the sides of Equations (2.7) and (2.8) gives

$$\gamma^2 V_x I_x = (R + j\omega L)(G + j\omega C) V_x I_x$$

so

$$\gamma = \sqrt{(R + j\omega L)(G + j\omega C)} \tag{2.10}$$

Again, for the ideal line, from (2.10)

$$\gamma = \sqrt{j^2 \omega^2 LC} = j\omega\sqrt{LC}$$

Since γ is purely imaginary

$$\alpha = 0$$

And the line has no attenuation.

$$\beta = \omega\sqrt{LC}$$

so

$$v_\mathrm{p} = \frac{\omega}{\beta} = \frac{1}{\sqrt{LC}}$$

and

$$v_\mathrm{g} = \frac{d\omega}{d\beta} \qquad \text{also} \qquad = \frac{1}{\sqrt{LC}}$$

L and C for a line are not independent – moving the lines apart makes L greater and C smaller. Theoretical calculations of L and C for a pair of parallel wires show that their product should be constant for a given medium surrounding the wires, and that $1/\sqrt{LC}$ is equal to the velocity of electromagnetic waves in the medium – 3×10^8 m/s if the wires are in a vacuum (see Appendix 2). This suggests that a more fundamental analysis of waves on a twin-wire line could be made by considering the electric and magnetic fields generated by the moving charges in the line.

For the general case where R and G are not zero, to get formulae for α and β the real and imaginary parts of the expression on the right hand side of Equation (2.10) must be worked out.

In the special, but unrealistic, *distortionless condition* in which $R/L = G/C$, Equation (2.10) can be rewritten

$$\gamma = \sqrt{LC(R/L + j\omega)(G/C + j\omega)} = (R/L + j\omega)\sqrt{LC}$$

yielding

$$\beta = \omega\sqrt{LC} \qquad \text{and} \qquad \alpha = R\sqrt{C/L}$$

which, since $R/L = G/C$, and for reasons which will become clear can be written

$$\alpha = \tfrac{1}{2}[R\sqrt{C/L} + G\sqrt{L/C}]$$

More generally, writing $\alpha + j\beta$ for γ in Equation (2.10) and squaring

$$(\alpha + j\beta)^2 = (R + j\omega L)(G + j\omega C)$$

Multiplying out brackets

$$\alpha^2 - \beta^2 + j2\alpha\beta = RG - \omega^2 LC + j\omega(LG + RC)$$

Equating real and imaginary parts

$$\alpha^2 - \beta^2 = RG - \omega^2 LC$$

$$2\alpha\beta = \omega(LG + RC)$$

By substituting from one into the other, these two simultaneous equations can be manipulated into a quadratic equation in α^2 and a quadratic equation in β^2, each of which can be solved in the normal way to yield the results:

$$\alpha = \{\tfrac{1}{2}[(R^2 + \omega^2 L^2)^{1/2}(G^2 + \omega^2 C^2)^{1/2} + (RG - \omega^2 LC)]\}^{1/2}$$

$$\beta = \{\tfrac{1}{2}[(R^2 + \omega^2 L^2)^{1/2}(G^2 + \omega^2 C^2)^{1/2} - (RG - \omega^2 LC)]\}^{1/2} \qquad (2.11)$$

Looking at Equation (2.11) it should be clear that in the general case neither ω/β nor $d\omega/d\beta$ are independent of frequency, thus both the phase velocity and the group velocity vary with frequency and they do not have the same value. However, if we multiply-out in Equation (2.10) and get

$$\gamma = \sqrt{RG + j\omega CR + j\omega GL - \omega'' LC}$$

when the frequency is high enough the term RG can be neglected and taking out of the bracket a term $(-\omega^2 LC)$ gives

$$\gamma = \sqrt{(-\omega^2 LC)\left(1 - j\frac{CR + GL}{\omega LC}\right)}$$

$$= j\omega\sqrt{LC}\left(1 - j\frac{CR + GL}{\omega LC}\right)^{1/2}$$

The second term in the second bracket, having ω in the denominator, will be small, so the binomial approximation can be applied (which is that $(1 + x)^n \approx (1 + nx)$ if $|nx| \ll 1$) to give

$$\gamma \approx j\omega\sqrt{LC}\left(1 - \frac{1}{2} j\frac{CR + GL}{\omega LC}\right)$$

Clearing the brackets

$$\gamma \approx \frac{1}{2}\frac{CR + GL}{\sqrt{LC}} + j\omega\sqrt{LC} \tag{2.12}$$

Equation (2.12) leads to the same results that were found for the distortionless condition: β is the same as for the loss-free line, so the phase and group velocities will be constant and equal.

$$\alpha = \tfrac{1}{2}[R\sqrt{C/L} + G\sqrt{L/C}]$$

which can be written

$$\alpha = \tfrac{1}{2}(R/Z_0 + GZ_0)$$

Since the characteristic impedance will be effectively resistive when this approximation holds, we can usefully write R_0 instead of Z_0 and G_0 for $1/Z_0$, giving a particularly neat equation

$$\alpha = \tfrac{1}{2}(R/R_0 + G/G_0)$$

R_0 being the characteristic resistance, $\sqrt{L/C}$, and G_0 the characteristic conductance, $\sqrt{C/L}$.

2.3   GROUP VELOCITY AND DISPERSION

The signals that we transmit in practice are never pure sine waves, but collections of sine waves which together constitute the spectrum of the signal; it is in this context that the notion of group velocity is significant. Consider a signal that consists of a single spike of voltage; an infinitely narrow pulse. The Fourier spectrum of such a pulse consists of an infinite set of sinusoids of all frequencies and equal amplitudes. If the propagating sine waves all travel at the same phase velocity they should cancel out everywhere except at one point giving a wave profile that is a pulse travelling with the same velocity as the phases of the waves.

But now, what if the component sine waves travel at different phase velocities? The pulse, which of course carries the energy, travels at a group velocity for which the value can be determined as follows.

Figure 2.6 shows the profiles of two adjacent sine waves from the infinite set. They differ in frequency and in phase velocity by an infinitesimally small amount, and hence they also differ in wavelength and in phase change coefficient by similar small amounts. Of necessity, the diagram vastly exaggerates the difference in wavelength.

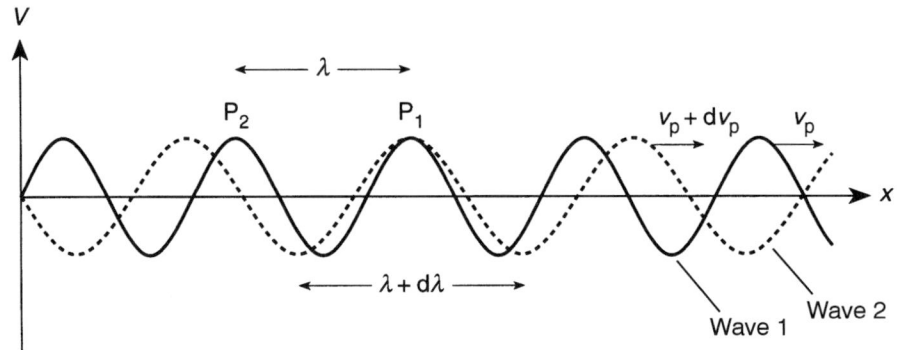

Fig. 2.6   Profiles of two adjacent sine waves in the spectrum of an infinitely narrow pulse

Wave 1 has:	angular frequency	ω
	phase velocity	v_p
	wavelength	λ
	phase change coefficient	β
Wave 2 has:	angular frequency	$\omega + d\omega$
	phase velocity	$v_p + dv_p$
	wavelength	$\lambda + d\lambda$
	phase change coefficient	$\beta + d\beta$

It may strike you that not all these increments can be positive together, but that will be taken care of by the mathematics.

At the instant shown, the two waves are at a peak together at the point P_1, which is where the pulse will be (when one takes account of the cancelling effect of all the other waves involved). Imagine that you, the observer, are travelling with wave 1, so that relative to you it is at rest. Wave 2 appears to be moving forward with a velocity dv_p, and after a time t two peaks coincide at P_2, so that the pulse position has moved back in wave 1 to position P_2 (all the other waves are altering their relative positions in similar fashion). Now

$$t = \frac{\text{relative distance travelled by wave 2}}{\text{relative velocity}} = \frac{d\lambda}{dv_p}$$

and in this time the pulse moves back relative to wave 1 a distance λ. Hence the relative velocity of the pulse is

$$-\lambda \div \frac{d\lambda}{dv_p} = -\lambda \frac{dv_p}{d\lambda}$$

The actual velocity of the pulse is this plus the velocity of wave 1, so

$$v_g = v_p - \lambda \frac{dv_p}{d\lambda} \tag{2.13}$$

Now $\lambda = 2\pi/\beta$ and $\lambda + d\lambda = 2\pi/(\beta + d\beta)$ so

$$d\lambda = \frac{2\pi}{\beta + d\beta} - \frac{2\pi}{\beta} = -\frac{2\pi \, d\beta}{\beta(\beta + d\beta)} \tag{2.14}$$

Also $v_p = \omega/\beta$ and $v_p + dv_p = (\omega + d\omega)/(\beta + d\beta)$ so

$$dv_p = \frac{\omega + d\omega}{\beta + d\beta} - \frac{\omega}{\beta} = \frac{\beta \, d\omega - \omega \, d\beta}{\beta(\beta + d\beta)} \tag{2.15}$$

From Equations (2.14) and (2.15),

$$\frac{dv_p}{d\lambda} = \frac{\omega \, d\beta - \beta \, d\omega}{2\pi \, d\beta} = \frac{1}{2\pi}\left(\omega - \frac{\beta \, d\omega}{d\beta}\right)$$

and substituting appropriately into Equation (2.13)

$$v_g = \frac{\omega}{\beta} - \frac{2\pi}{\beta} \cdot \frac{1}{2\pi}\left(\omega - \frac{\beta \, d\omega}{d\beta}\right) = \frac{d\omega}{d\beta}$$

This result will apply equally well to any sort of pulse. If a pulse is to travel along without changing its shape, then the group velocity associated with different parts of its spectrum must be the same; in other words, $d\omega/d\beta$ must be independent of frequency. If it is not, then the pulse will change its shape as it travels, generally giving a waveform at the receiver that is spread out in time: this effect is called *dispersion*. The amount of dispersion is usually expressed as the time spread in the received pulse, between half-power points, for a transmitted infinitely narrow pulse.

If a graph is drawn of β against ω (β is the natural dependent variable), then the slope of this curve is $d\beta/d\omega$, the inverse of group velocity. This quantity is known as the *group delay*. If the curve is a straight line through the origin, then the phase velocity and group velocity will be the same at all points and equal.

Figure 2.7 shows the β/ω curve for a practical line. At ω_1, v_p is the inverse slope of OA, v_g is the inverse slope of PQ. Similarly at other frequencies. Clearly, signals with low frequency components will suffer dispersion.

In general, if a transmission medium has a phase velocity that changes with frequency, the group velocity differs from the phase velocity and also varies with frequency so that the medium is dispersive.

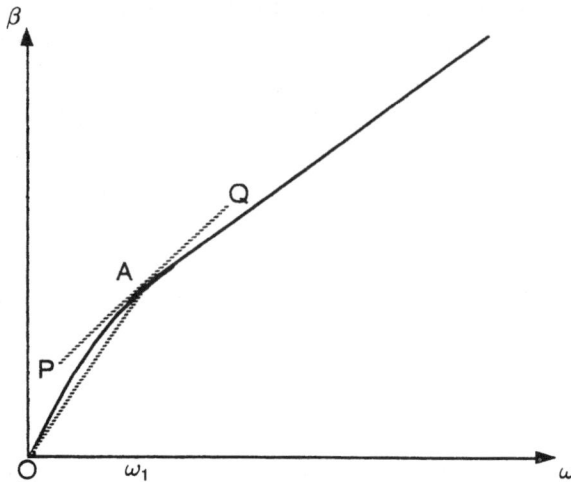

Fig. 2.7 β/ω curve for a practical line

2.4 REFLECTIONS

Unless specific adjustments are made to avoid it, a wave that reaches the end of a transmission line will be reflected back in part, with undesirable consequences.

Two-way propagation of pulses

First we discuss the propagation of two pulses going in opposite directions on an ideal transmission line (no dispersion) and passing through each other. The voltage profile of the two waves on the line, before they meet, is shown in Fig. 2.8(a).

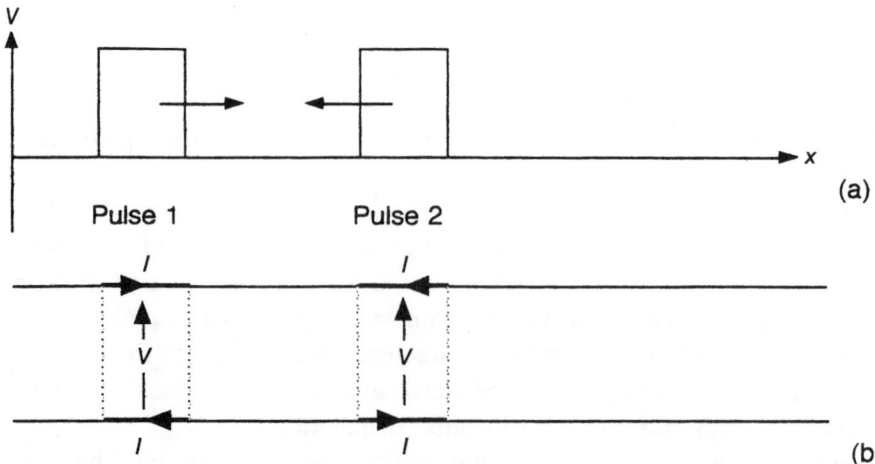

Fig. 2.8 Two pulses of the same polarity and amplitude travelling in opposite directions

The distribution of voltages and currents in the wires is shown in Fig. 2.8(b). Since pulse 1 is moving to the right, with the indicated voltage polarity, the current in the wires will be as shown: this is consistent with the power being delivered from left to right. Pulse 2, which is taken to be of the same voltage amplitude, is moving to the left, and so the current directions are opposite to those in pulse 1. In each pulse, $I = V/R_0$ where R_0 is the characteristic impedance of the line (resistive, since the line is ideal).

It is worth stopping and thinking a little more about what is happening in a pulse. The current in the pulse represents the moving of electrons into and out of the wires of the next part of the line, thinning out to cause a net positive charge and crowding together to form a net negative charge, to set up the voltage in the next part of the line. Over the extent of the pulse there is an electric field between the charges in the wires accounting for the voltage between them, and there is a magnetic field associated with the currents. The wave (that is, the propagating pulse) can be described either in terms of currents and voltages, or in terms of moving charges, or in terms of electric and magnetic fields: detailed study of electromagnetic theory suggests that the energy is actually stored and propagated in the changing fields.

When the two pulses meet they will pass through each other and while they are doing so, in the region of overlap the voltage will be double that in either pulse and the net current will be zero – *see* Fig. 2.9.

Suppose now that pulse 2 were inverted. As they passed through each other the net voltage in the overlap would be zero and the net current double – *see* Fig. 2.10.

In general, if the two pulses are not the same amplitude, then if they are the same way up in terms of voltage, as they pass through each other their voltages

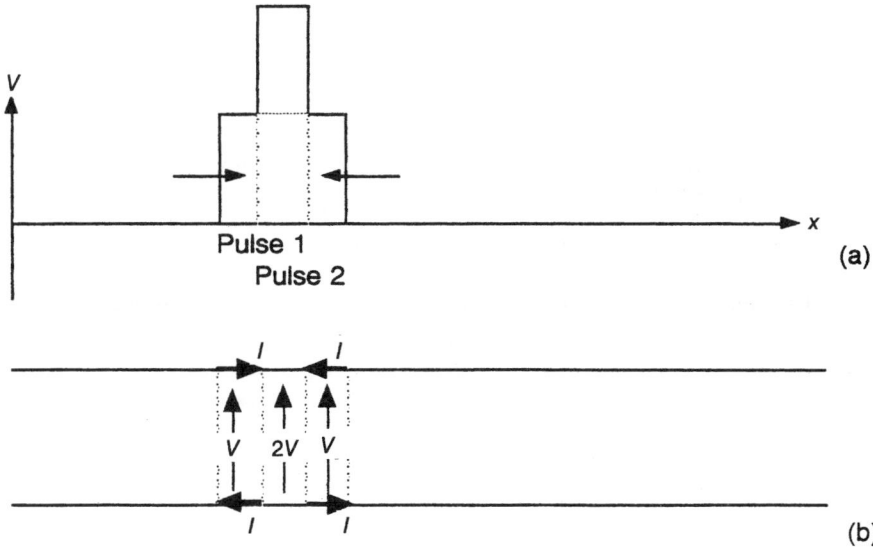

Fig. 2.9 Two pulses of the same polarity and amplitude travelling in opposite directions on a line passing through each other

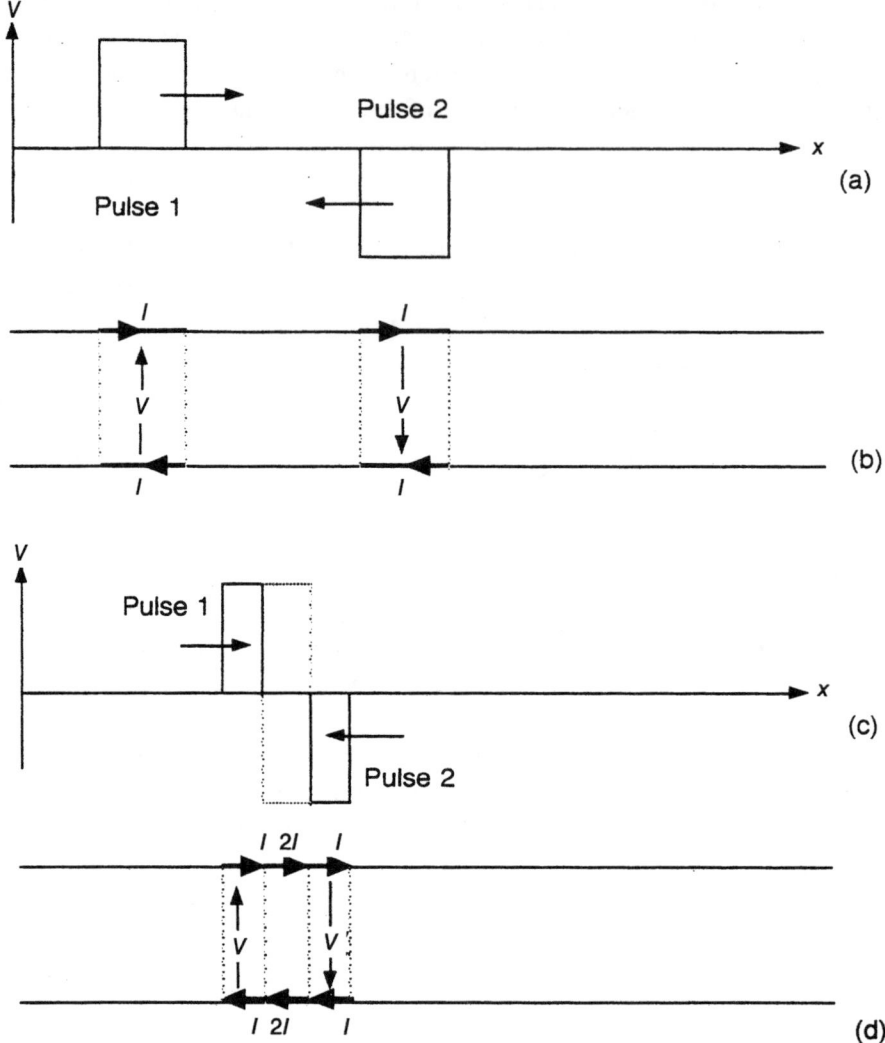

Fig. 2.10 Two pulses of the same amplitude and opposite polarity travelling in opposite directions on a line passing through each other

will add and their currents subtract in the overlap, whereas if they have opposite voltage polarities their voltages will subtract and their currents add.

Reflection of pulses

Suppose that at the receiving end of a line the two wires are connected together producing a short circuit. A pulse, driven on to the line at the driving point, will travel along until it meets the short circuit: what happens then? The energy in the pulse has to go somewhere, and the short circuit cannot absorb it because it has no resistance.

The clue lies in the previous section. When two pulses of equal magnitude, opposite voltage and travelling in opposite directions overlap, the resultant voltage is zero, while a current flows which is twice that associated with either pulse. At the short circuit, current will flow with zero voltage, so that the subsequent propagation must be two equal and opposite pulses, one moving into the short circuit and the other moving out of it back along the line. The effect is reflection with inversion.

If the end of the line is simply left disconnected the termination is an open circuit and can have a voltage but zero current. A similar argument to the previous one should lead to the conclusion that this time the pulse will be reflected without inversion. This case is illustrated in Fig. 2.11.

Is there any way of dealing with the end of a transmission line so that no reflection occurs? If a resistance equal to the characteristic impedance of the line (remember, this is a line without dispersion, so its characteristic impedance will be resistive) is connected to the end then its ratio of current to voltage is exactly right to absorb the energy from the pulse as it arrives. Looking at it another way, a further infinite length of line, connected to the end of the actual line, would have an input impedance R_0; the pulse, when it reaches the termination, could as well be entering a further section of line. Of course, one does not simply want to absorb the pulse in a resistor – presumably it has been sent for a purpose – but the input impedance of any device at the receiving end must be R_0 to prevent reflection: under these circumstances the line is said to be *correctly terminated* or *matched*.

If the terminating impedance is not R_0 there will be partial reflection; some of the energy will be absorbed and some will not. When the reflected wave arrives back at the driving point it will be reflected again unless the generator has an internal impedance R_0. Multiple reflections can cause all sorts of problems in transmission systems and should be avoided as far as possible.

The size and polarity of the reflected pulse can be calculated if you know the value of the terminating resistance R_L. Look at Fig. 2.11. Since the voltage in the pulse *is not* inverted on reflection, the current in the pulse *is* inverted (if the voltage were inverted the current would not be).

total voltage $= V_i + V_r$

total current $= I_i - I_r$

Their ratio must be R_L, so

$$\frac{V_i + V_r}{I_i - I_r} = R_L \tag{2.16}$$

It proves useful to take the ratio V_r/V_i: this ratio is called the voltage *reflection coefficient* and given the symbol ρ.

Since, taking account of the polarities given to the currents and voltages in Fig. 2.11

$$V_i/I_i = V_r/I_r = R_0$$

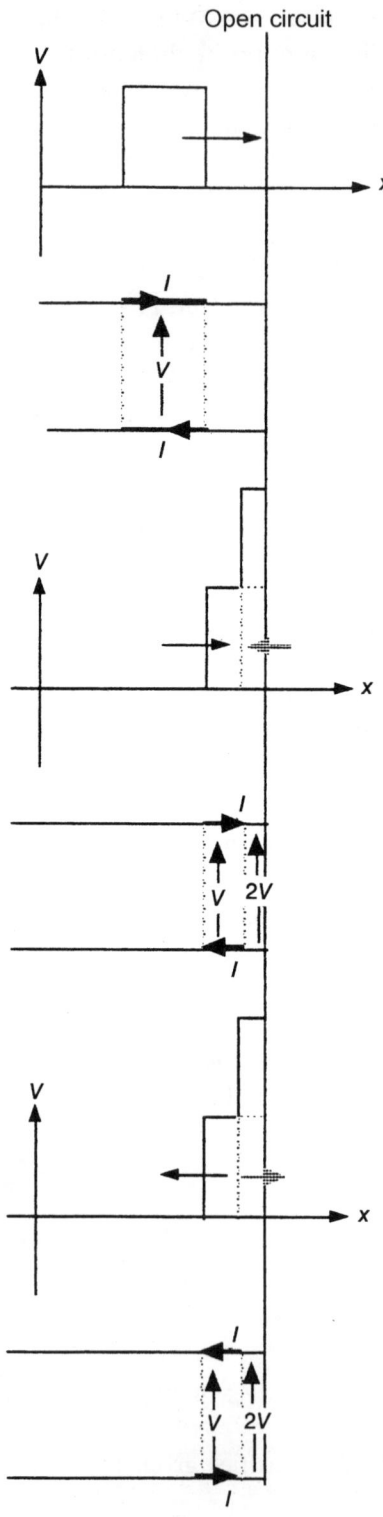

Fig. 2.11   A pulse being reflected from an open circuit

it follows that $\rho = I_r/I_i$ as well as V_r/V_i. (Notice that the current ratio is taken as positive when the two currents are in *opposite* senses, whereas the voltage ratio is positive when the two voltages are in *the same* sense.)

Equation (2.16) can be rewritten

$$R_L = \frac{V_i}{I_i} \frac{1 + V_r/V_i}{1 - I_r/I_i}$$

so

$$R_L = R_0 \frac{1 + \rho}{1 - \rho}$$

which by algebraic manipulation gives

$$\rho = \frac{R_L - R_0}{R_L + R_0}$$

Putting numbers in to give two examples:

Suppose $R_L = 2R_0 \qquad \rho = 1/3$

the reflected pulse magnitude will be 1/3 of the incident pulse magnitude.

Now let $R_L = \frac{1}{2}R_0 \qquad \rho = -1/3$

the minus sign tells us that there is voltage inversion, so the reflected voltage magnitude of the pulse is 1/3 that of the incident pulse but inverted.

In general, if the terminating resistance is greater than the characteristic impedance, the voltage of a reflected pulse is not inverted; if less, then it is inverted. Open and short circuit loads are limiting cases already discussed.

The ratio of energy in reflected pulses to energy in incident pulses is called the *return loss* and is generally expressed in dB. Energy is proportional to voltage squared, so at the load the return loss equals

$$10 \log_{10}(\rho^2) \qquad \text{or} \qquad 20 \log_{10}|\rho| \, \text{dB}$$

For both cases considered above the return loss at the load equals $-9.5\,\text{dB}$.

Reflection of sine waves

A sine wave will be reflected from a resistive load in a similar way to that described for pulses: if the load resistance is greater than the characteristic impedance then the voltage wave will be reflected so that the phase of the reflected wave is the same as that of the incident wave at the load; if the load is less than the characteristic impedance then the phase is inverted on reflection.

If the load is not purely resistive but has a reactive component (inductive or capacitive) then reflection involves a phase change other than $180°$. A repeat of the argument used in the previous section, but considering the phasor voltages and currents in the sine wave, leads to

$$\rho = \frac{Z_L - Z_0}{Z_L + Z_0}$$

where ρ is now a complex reflection coefficient defined as the phasor ratio V_r/V_i and I_r/I_i. It follows that if a pulse is reflected by a non-resistive termination, even if the line is non-dispersive the reflected pulse will be distorted because the sinusoidal components of the pulse will suffer different phase shifts on reflection.

For a single sine wave transmitted and reflected, the resultant of the two sine waves travelling in opposite directions is called a *standing wave*. If the reflected wave is of smaller amplitude than the incident wave we have a *partial standing wave*: the resultant when the waves are of equal amplitude is a *total standing wave*.

A total standing wave

In the analysis which follows we shall assume that attenuation is negligible over the lengths of line considered. Look at Fig. 2.12. A sine wave is travelling from the left towards a termination which is an open circuit. The distance from a chosen point, marked P, to the termination is a.

If the voltage in the incident wave, at the point P, is represented by the phasor V, then, at the termination, the incident-wave phasor will be $Ve^{-\gamma a}$, and if the attenuation is negligible, $\gamma = j\beta$, so the phasor voltage at the point of reflection is $Ve^{-j\beta a}$. I shall call this phasor voltage V'. This is also the phasor voltage of the reflected wave at the termination, so the phasor voltage of the reflected wave at P, taking account of the phase lag back to P in the returning wave, must be $V'e^{-j\beta a}$.

Since $V' = Ve^{-j\beta a}$, then $V = V'e^{j\beta a}$, so the total voltage at P due to both the incident and reflected waves can be written

$$V_P = V'e^{j\beta a} + V'e^{-j\beta a} = V'(e^{j\beta a} + e^{-j\beta a})$$

It can be shown mathematically that

$$(e^{j\beta a} + e^{-j\beta a}) = 2\cos(\beta a)$$

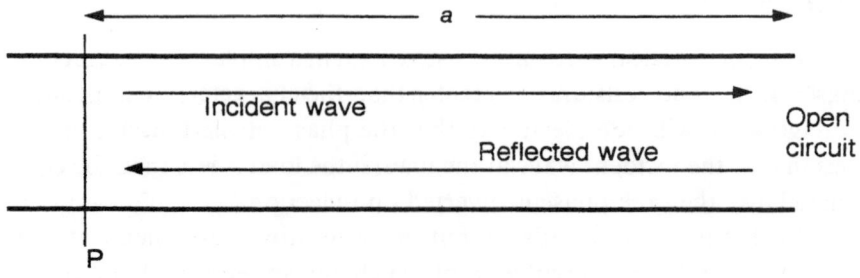

Fig. 2.12   Incident and reflected waves on an open circuit line

and it has already been shown that $\beta = 2\pi/\lambda$ so

$$V_P = 2V' \cos(2\pi a/\lambda)$$

Since the cosine of 0 is 1, at the termination, where $a = 0$, $V_T = 2V'$, so finally one can write

$$V_P = V_T \cos(2\pi a/\lambda) \tag{2.17}$$

If you consider the term $\cos(2\pi a/\lambda)$ as a increases, it varies from 1 when $a = 0$ to 0 when $2\pi a/\lambda = \pi/2$, i.e. when $a = \lambda/4$. It then increases in magnitude negatively up to -1 when $a = \lambda/2$. Continuing in this way, it can be seen that V_P varies in magnitude between 0 and the magnitude of V_T in a cyclic way. At all points where $4a/\lambda$ is an odd integer V_P is zero, and at all points where $4a/\lambda$ is an even integer V_P has maximum amplitude. Points on the line at which V_P has maximum amplitude are called *antinodes*; points at which V_P is zero are called *nodes*. Nodes are separated by half a wavelength and antinodes the same: nodes and antinodes alternate along the line at quarter-wave intervals as shown in Fig. 2.13.

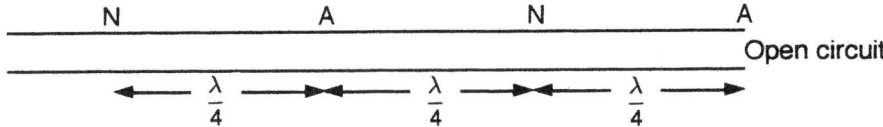

Fig. 2.13 Nodes and antinodes on a total standing wave on an open circuit line

It can be seen from Equation (2.17) that the total standing wave does not have a continuous phase variation with distance: the phase of the voltage is constant over the distance between two nodes (it inverts on passing through a node). A physical explanation is that as we move nearer to the load the phase of the forward wave moves backwards and the phase of the returning wave moves forwards so that when we add the two together the phase shifts cancel out.

An analysis in terms of current indicates a similar standing wave of current, with nodes alternating with antinodes at quarter wave intervals, but current antinodes occur at voltage nodes and vice versa. At each voltage antinode (current node) the impedance of the line – the ratio V_P/I_P – in the 'steady state' (that is, when the incident and reflected voltage are both established on the line) is infinite: at voltage nodes the impedance is zero. At other points on the line the impedance is a pure reactance (either inductive or capacitive depending on the position); this must be so since the average power flow into the line must be zero.

This discussion started with a line terminated with an open circuit – giving a voltage antinode at the termination; if the line were terminated with a short circuit the same total standing wave pattern would result, but with a voltage node at the termination. Terminating the line with any pure reactance would again result in no absorption of energy, and so a total standing wave pattern; the termination would not coincide with a node or antinode, but the position of the pattern would be such as to result in the correct impedance at the termination.

A partial standing wave

If only part of the wave energy is reflected back, a partial standing wave will result. If the incident wave is thought of as the sum of two parts, one of which forms with the reflected wave a total standing wave, it will be seen that a partial standing wave can be modelled as a total standing wave with a travelling wave superimposed. Again there is a pattern of maxima and minima; the voltage maxima are the *sum* of the incident and reflected voltage amplitudes, the minima are their *difference* (no longer zero). Similarly for the current maxima and minima. Relative positions are as for the total standing wave. The resultant phase now varies with distance, but not linearly; phase changes more rapidly in the region of a minimum.

The ratio of the maximum voltage amplitude to the minimum voltage amplitude (a quarter wavelength away) is called the *voltage standing wave ratio* (VSWR).[1] In many practical circumstances this parameter can be readily measured, and it may be used as an indicator of the quality of termination of the line.

The voltage standing wave ratio can be related to the modulus (magnitude) of the reflection coefficient. Call the amplitude of the incident wave V_i and that of the reflected wave V_r. At a voltage maximum in the standing wave[2]

$$V_{max} = V_i + V_r$$

and at a minimum

$$V_{min} = V_i - V_r$$

so

$$S = \frac{V_{max}}{V_{min}} = \frac{V_i + V_r}{V_i - V_r}$$

Dividing top and bottom by V_i

$$S = \frac{1 + V_r/V_i}{1 - V_r/V_i}$$

V_r/V_i is the modulus of the phasor ratio $\boldsymbol{V_r}/\boldsymbol{V_i}$ at the termination (or anywhere else), so it is the modulus of the reflection coefficient $|\rho|$, hence

$$S = \frac{1 + |\rho|}{1 - |\rho|}$$

Since usually the standing wave ratio is easier to measure than the reflection coefficient, this gives a way of deducing $|\rho|$. The formula can be manipulated to give

$$|\rho| = \frac{S - 1}{S + 1}$$

[1] The VSWR is only a meaningful measurement if the attenuation per wavelength is small.
[2] In formulae, the single letter S is used for VSWR

A standard sometimes applied to high frequency line terminations is that the VSWR should not exceed 1.5:1. This gives

$$|\rho| = \frac{0.5}{2.5} = \frac{1}{5} = 0.2$$

$0.2^2 = 0.04$ so 4 per cent of the power is reflected

return loss $= 20 \log_{10} 0.2 = -14\,\text{dB}$

The following example illustrates the sort of measurement which is often carried out on a line and what can be deduced from it.

Example calculation

A high frequency transmission line, of characteristic impedance $50\,\Omega$, is terminated by a transmitting antenna. With a sine wave signal on the line the voltage standing wave ratio is found to be 1.4:1 and one of the voltage minima is located at exactly $2\frac{1}{4}$ wavelengths from the antenna terminals. From these data it is possible to deduce the nature of the load presented by the antenna to the line as follows.

Voltage minima are $\lambda/2$ apart and interleaved with voltage maxima, so at $2\frac{1}{4}$ wavelengths from a voltage minimum is a voltage maximum. The sine wave is thus reflected without change of phase by the antenna; ρ is real and positive and so the load is resistive and greater than R_0.

$$|\rho| = \frac{S-1}{S+1}$$

so in this case

$$\rho = \frac{0.4}{2.4} = \frac{1}{6}$$

Now

$$\rho = \frac{R_\text{L} - R_0}{R_\text{L} + R_0}$$

so again, in this case

$$\frac{1}{6} = \frac{R_\text{A} - 50}{R_\text{A} + 50} \qquad \text{yielding} \quad R_\text{A} = 70\,\Omega$$

The antenna presents a resistive load of $70\,\Omega$ to the line.

If the termination has a reactive component, the distance from a conveniently located voltage minimum will not be an exact number of quarter-wavelengths and the calculation will be more complex. A special graphical technique using a *Smith chart* facilitates such calculations, *see* Chapter 7.

Resonant line-sections

If a short piece of transmission line, short circuited at the load-end, is fed with a sine wave of a frequency for which it is exactly $\frac{1}{4}$ wavelength long, the input impedance will be infinite (or very large if one allows for finite attenuation). A small drop in frequency makes the wavelength longer and so the line-section less than $\lambda/4$ so that the input impedance is inductive; raising the frequency a little makes the section capacitive. These are similar to the properties of a parallel tuned circuit. It is difficult to make components which behave as reasonably pure inductors or capacitors at higher frequencies – say above 500 MHz – so resonant line sections are often used in this way (*see* Chapter 7).

Another common example is a quarter-wave section of line, of intermediate characteristic impedance, introduced between the main line and a resistive load to effect matching (but only at the appropriate frequency). The required characteristic impedance of the matching section proves to be

$$Z_\mathrm{m} = \sqrt{Z_0 Z_\mathrm{L}}$$

(*see* Chapter 7).

Equalization

The variation of attenuation with frequency and the dispersion of a transmission line can be compensated for to some extent by connecting to the end of the line before the load an equalizer circuit. This is a type of filter circuit which is tailored to have an amplitude/frequency characteristic and a phase-shift/frequency characteristic which are complementary, as far as possible, to those of the line.

Effects of reflections

We have seen that if the load is not matched to the line some of the power is reflected back – and hence wasted. This is not the only reason why reflection has to be controlled.

It has already been suggested that multiple reflections from the two ends of the line must be avoided; they can cause erroneous signals to be received.

Reflections can also be caused by discontinuities in the line – particularly at joints. These need to be minimized.

If high powers are in use – as in the case of a feeder from a transmitter to an antenna – voltage maxima caused by reflection can lead to electrical breakdown.

Finally, reflected power arriving back at the input may affect the generator, causing both its frequency and power output to vary in an unpredictable manner.

Fault location using pulses

Pulses can be used to find the position of a fault on a line, and some clue as to the nature of the fault. The following example illustrates this.

Example calculation
A correctly terminated transmission line of characteristic impedance $100\,\Omega$ has a measured velocity of propagation of certain pulses of 2.2×10^8 m/s and an

attenuation, for those pulses, of 5 dB/km. A fault develops, and pulses sent down the line are reflected so that they arrive back at the input inverted, with a magnitude 0.043 of that of the transmitted pulses, and with a time lapse of 21 μs. From these data the position and nature of the fault can be estimated as follows.

The pulses must take $21/2$ μs to get to the fault, so its position is

$$(2.2 \times 10^8) \times (10.5 \times 10^{-6})\,\text{m} = 2.31\,\text{km}$$

from the sending end. Since the pulses are inverted, the impedance at the point of reflection must be less than the characteristic impedance, and since the rest of the line is matched and so presents an impedance of R_0 at the fault, the fault condition must be a partial short circuit – *see* Fig. 2.14(a), which can be redrawn as in Fig. 2.14(b). The magnitude of the reflection coefficient at the fault is

$$\frac{R_0 - R_{\text{Fault}}}{R_0 + R_{\text{Fault}}} \quad \text{and} \quad R_{\text{Fault}} = \frac{R_{\text{sc}} \times R_0}{R_{\text{sc}} + R_0}$$

giving

$$|\rho| = \frac{R_0^2}{R_0^2 + 2R_{\text{sc}}R_0} = \frac{10^4}{10^4 + 200R_{\text{sc}}}$$

The fall in amplitude of the pulse due to the round trip on the line is

$$2 \times 2.31 \times 5\,\text{dB} = 23.1\,\text{dB}$$

which is a voltage ratio of $10^{23.1/20} = 14.3$. So,

$$\frac{1}{14.3} \times |\rho| = 0.043$$

which yields $R_{\text{sc}} \approx 31\,\Omega$

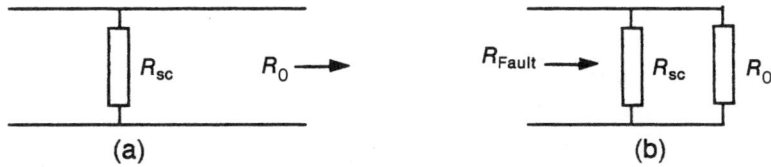

(a)   (b)

Fig. 2.14   Representation of a short circuit fault

The above is an example of a general technique for finding line faults, known as *time domain reflectometry*, that is widely used on both line systems and optical fibre systems. In some cases a step function is transmitted rather than a pulse. Such a step can be thought of as the front of a very long pulse; the effect is similar, in that a smaller version of the step arrives back which may or may not be inverted.

In practice there will always be a number of small reflections on a line: the termination will not be perfect, and every joint on the line will cause some reflection. When a line is working satisfactorily it is possible to 'map' the reflection pattern; significant deviations from this can then be used to diagnose deteriorations in the line's performance.

Padding and isolation

The complete avoidance of reflection is not possible: if the attenuation of the line is sufficient then this may reduce the reflected signal to an acceptable level before it reaches the input; if not then it may be necessary to add some lumped attenuation. This is known as *padding*.

Padding, of course, reduces the outgoing signal as well as the reflected signal: in transmission media which are used at the highest frequencies passive devices are available which attenuate the return signal but not the outgoing signal. These devices, known as *isolators*, are not available in twin-wire line, but the same effect can be achieved using an (active) amplifier as a 'buffer'.

2.5   NOISE AND INTERFERENCE

In any transmission system a parameter of primary importance is the signal to noise ratio at the receiver output. Some of the noise is generated in the receiver input circuits; that noise which comes from the line is now discussed.

A detailed analysis involving statistical mechanics and quantum theory (two branches of theoretical physics) shows that in any matched transmission system, due to the effects of heat in generating random transient electrical disturbances, the medium delivers to the receiver a fundamental noise power given by the formula

$$P_N = kTB \left[\frac{hf/kT}{\exp(hf/kT) - 1} \right] \tag{2.18}$$

where k is Boltzmann's constant and h is Planck's constant. T is the absolute temperature of the medium, B is the receiver bandwidth and f is the receiver mid-band frequency. (The expression hf/kT, in fact, compares the size of the quantum of radiation at frequency f to the average heat energy per constituent particle of the medium at temperature T, so the term in square brackets would require to be integrated over the frequency range for a very wide percentage bandwidth.)

At room temperature and all frequencies below about 10^{12} Hz, effectively the term in brackets equals 1, so

$$P_N = kTB$$

At optical frequencies, however, the formula yields a noise power much lower than kTB. This will be discussed further in Chapter 5.

The transmitter circuits may generate some noise, but in all practical systems the signal to noise ratio at the transmitter is high. Both the transmitter signal and the transmitter noise are attenuated to the same extent in transmission, so that, at the receiver input of a matched transmission line system, the transmitter noise is negligible compared to the line noise and the signal to fundamental noise ratio is effectively

$$\frac{P_s}{kTB}$$

where P_s is the received signal power.

This basic signal to noise ratio will be degraded by any extraneous voltages coupled onto the line: these can come from two sources, interference and cross-talk. Interference comes from voltages induced into the line by such events as electrical equipment switching on or off; it can be avoided by carefully screening the line with an earthed conducting sheath. Cross-talk occurs because lines often have to be run side by side in the same cable. There is coupling – mainly capacitive, but it can also have an inductive component – between wires of different lines. To minimize this, the wires of a line are twisted round each other so that first one wire and then the other is in proximity with a wire of another line and induced voltages cancel. Since all the lines in a cable will be twisted, they are twisted at different pitches (number of twists per metre) so that the cancelling can occur.

2.6 SUMMARY OF GENERAL WAVE PROPAGATION

These are the ideas that will be made use of in subsequent chapters.

For any travelling sinusoidal wave there is a progressive increase in phase delay and a progressive reduction in amplitude with distance in the direction of propagation. These are characterized by a phase change coefficient β and an attenuation coefficient α respectively.

The profile of a sinusoidal wave travels at the phase velocity, v_p, which is equal to $f\lambda$ or ω/β.

A pulse of wave energy travels at the group velocity, v_g, which is equal to $d\omega/d\beta$.

If v_g is independent of frequency the medium is non-dispersive; this is, in particular, the case when $v_g = v_p$.

The reflection of pulses at the termination of a confining medium (not free space) can be characterized by a reflection coefficient, ρ, which is the ratio of the amplitude of a reflected pulse to that of an incident pulse. The proportion of pulse energy reflected back to any point is the return loss; at the termination this is given by $20 \log_{10} |\rho|$ dB.

2.7 SUMMARY OF SPECIFIC PROPERTIES OF LINE PROPAGATION

For a travelling sinusoidal wave on a transmission line:

The currents and voltages can be represented by phasors $Ve^{-\gamma x}$ and $Ie^{-\gamma x}$, where γ, the propagation constant, contains both amplitude and phase information; $\gamma = \alpha + j\beta$.

The ratio of phasor voltage to phasor current at a point has, at a given frequency, a constant value for all points and is called the characteristic impedance, Z_0.

In terms of the line primary constants:

$$\gamma = \sqrt{(R + j\omega L)(G + j\omega C)}$$

$$Z_0 = \sqrt{\frac{R + j\omega L}{G + j\omega C}}$$

At frequencies where $\omega L \gg R$ and $\omega C \gg G$ (including the loss-free line as a special case):

the attenuation per wavelength is small.

$Z_0 \approx \sqrt{L/C}$ and is resistive (purely real).

$\beta \approx \omega\sqrt{LC}$ and so $v_p \approx 1/\sqrt{LC}$ – which is independent of frequency, so the line is non-dispersive.

$\alpha \approx \frac{1}{2}(R/R_0 + G/G_0)$

A complex reflection coefficient can be defined as

$$\rho = \frac{\text{Phasor voltage of reflected wave}}{\text{Phasor voltage of incident wave}}$$

In terms of impedances

$$\rho = \frac{Z_L - Z_0}{Z_L + Z_0} \qquad \text{or rearranging} \qquad Z_L = Z_0\frac{1 + \rho}{1 - \rho}$$

Reflection of a sine wave produces a standing wave with a **VSWR** value S; this is related to the magnitude of the reflection coefficient by

$$|\rho| = \frac{S - 1}{S + 1}$$

2.8   LINE CALCULATIONS

The following information refers to a particular grade of *data transmission cable*, and is in a form in which line characteristics are often quoted:

Characteristic impedance 78 Ω
Capacitance 65 pF/m

Attenuation (per 100 m)	1 MHz	10 MHz	50 MHz
	2 dB	6.9 dB	16.4 dB

2.1 What is the inductance per metre of this line?

2.2 Calculate the value of the attenuation coefficient at 1 MHz.

2.3 Estimate, at a frequency of 1 MHz, the value of R if G were negligible and the value of G if R were negligible and so, by comparing ωL with R and ωC with G, show that the line is non-dispersive at this frequency.

2.4 Show how the given data indicate that α increases with frequency and hence show that the line will be non-dispersive at frequencies above 1 MHz.

2.5 Calculate the velocity of propagation on the line at the three quoted frequencies.

2.6 Calculate the wavelength on the line at 10 MHz.

2.7 What would be the phase difference in degrees between two points on the line 10 metres apart at a frequency of 50 MHz?

A 200 m length of the line is used to connect a transmitter, generating pulses of 50 mHz radiation, to a receiver with an input impedance consisting of a resistance of 100 Ω in parallel with a capacitance of 20 pF.

2.8 Calculate the magnitude of the reflection coefficient at the receiver.

2.9 Work out the return loss at the transmitter and hence the ratio of the reflected pulse amplitude to the transmitted pulse amplitude at the transmitter.

3 Electromagnetic waves

An understanding of propagation in the other media with which this book is concerned requires an understanding of the nature and properties of electromagnetic waves. The discussion will be mainly in terms of classical electromagnetism; however, it must be understood that this theory is incomplete, in that it does not account for the quantization of the energy in electromagnetic radiation, and some reference will have to be made to photons in the context of radiation at optical frequencies.

In this chapter electromagnetic waves are simply described, without proving that they have the properties detailed: some analysis, starting from Maxwell's equations, will be found in Appendix 4.

Classical electromagnetism involves the following relevant parameters:

Electric field strength, E (SI units volts per metre)
Magnetic field strength, H (SI units amperes per metre)
Permeability of free space, μ_0 (value $4\pi \times 10^{-7}$ henries per metre)
Permittivity of free space, ε_0 (value 8.854×10^{-12} farads per metre)
Relative permittivity (dielectric constant), ε_r
Total permittivity (or, simply, 'permittivity'), ε ($\varepsilon = \varepsilon_r \varepsilon_0$)

The discussion will be concerned with media which are isotropic (that is, have the same properties irrespective of direction), non-conducting, non-magnetic and loss-free (or almost so). Free space is the obvious example, but air, a number of dielectric materials, silica and other forms of glass, can all approximate to these conditions over some frequency ranges.

Electric fields can terminate on charges and are associated with voltages between conductors; magnetic fields are associated with currents in conductors. However, even when no conductors are present, varying electric and magnetic fields can exist together because a varying electric field produces a magnetic field and a varying magnetic field produces an electric field. The consequence is that electric and magnetic field variations can transport energy as an electromagnetic wave with sheets of varying electric field generating sheets of varying magnetic field and the sheets of varying magnetic field generating sheets of varying electric field and so on.

3.1 ELECTROMAGNETIC PLANE WAVES

All time variations of fields will be taken as sinusoidal. x, y and z will be used as three-dimensional coordinates as indicated in Fig. 3.1.

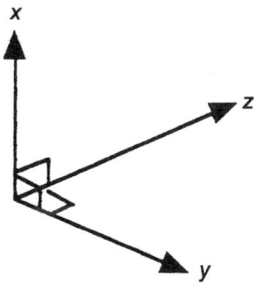

Fig. 3.1 Choice of three-dimensional coordinates

The *linearly polarized plane wave* is the simplest form of electromagnetic wave; even so it is difficult to draw a diagram that will adequately illustrate it – an attempt is shown in Fig. 3.2. All the electric field points in one direction; the x direction has been chosen in the diagram. The magnetic field is in phase with the electric field at every point (both vary together in exactly the same way) and points in a direction at right angles to the electric field – the y direction in the diagram. The diagram can only catch the electric and magnetic field *maxima* over a small area at a given instant of time. One has to imagine that

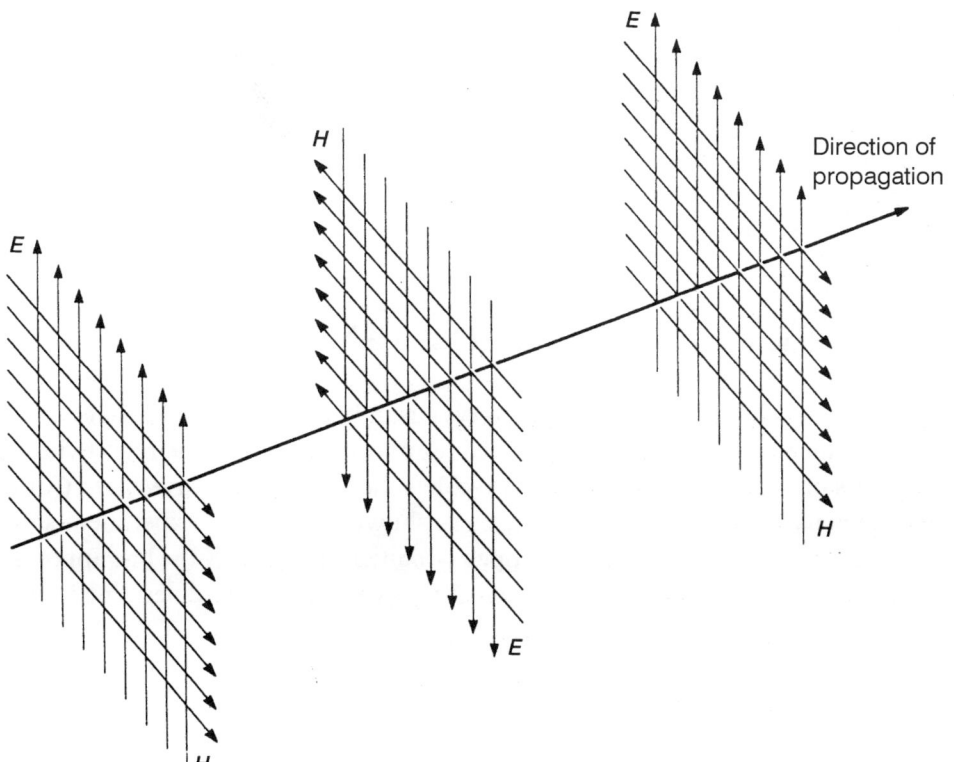

Fig. 3.2 Electric and magnetic field maxima at one instant in an electromagnetic plane wave

the fields extend to infinity in the x and y directions and that between these planes of electric and magnetic field maxima there are planes containing electric and magnetic fields which, because they happen to be at a different phase of the sinusoidal variation, have, at this instant, lower field values. But of course as the wave progresses each plane will become, in turn, the plane in which the electric and magnetic fields are maxima, and so the whole pattern travels forwards in the direction shown as the direction of propagation.

Mathematical analysis (Appendix 4) indicates that a loss-free wave of electric and magnetic field propagates in the z direction with a phase velocity $1/\sqrt{\mu_0 \varepsilon}$. For free space this is $1/\sqrt{\mu_0 \varepsilon_0}$ which has a value $\approx 3 \times 10^8$ m/s and is given the symbol c. The ratio $E/H = \sqrt{\mu_0 / \varepsilon}$: this ratio has the dimensions ohms and is called the *wave impedance*; its value for free space is 120π, i.e. $377\,\Omega$.

The product $E \times H$ has the dimensions volts/metre times amps/metre, giving watts/metre^2 and indicates the rate of energy flux through 1 square metre of area perpendicular to the direction of propagation – that is, the power flux density. E and H are varying together sinusoidally, so that the instantaneous power flux density at a given point is constantly changing; however, the significant parameter is the *average* power flux density, $E_{rms} \times H_{rms}$, which is often simply described as the *power density* of the wave.[1]

This plane wave is said to have a direction of *polarization*, which tells us how it is oriented in space, and which, by convention, is taken as the direction of the electric field – i.e. the x direction in the coordinates used here.

It is difficult to see how such a wave could be launched: rather than a driving point one would need a driving plane over which one would need to impose the necessary uniform varying electric and magnetic fields at right angles. In fact the plane wave represents a particularly simple form to which other wave distributions can be converted or considered to approximate.

3.2   SPHERICAL ELECTROMAGNETIC WAVES

At the driving point, electromagnetic waves usually start out as spherical waves: a simple launching device is now described. Suppose that an open-circuit twin-wire transmission line, driven with a sine wave, has the last quarter wavelength (for the frequency used) bent out at right angles as shown in Fig. 3.3. Experiment shows that the line no longer appears to be open circuit: at the point X the line appears to be terminated by a resistance of about $70\,\Omega$ (a value largely determined by the properties of space and not by the characteristic impedance of the line). The energy is not converted into heat, but is radiated away; the termination is called a *half-wave dipole*.

---

[1] The electric and magnetic fields are vectors and, in general, the vector cross product $\boldsymbol{E} \times \boldsymbol{H}$ gives the power flux density: this is Poynting's theorem. Since $\boldsymbol{E}$ and $\boldsymbol{H}$ are at right angles in this case, the result is simply the product of the two magnitudes.

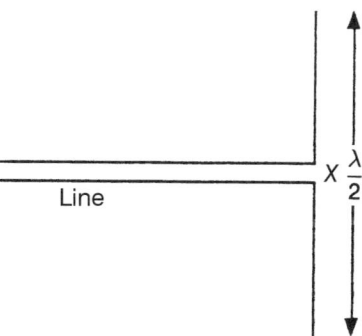

Fig. 3.3 A line bent out to form a half-wave dipole

Along the dipole there are sinusoidally varying currents and voltages which create fields around it. Near to the dipole the distribution of fields is complicated, but beyond two or three wavelengths away emerge electric and magnetic fields in phase and at right angles to each other which propagate as an electromagnetic wave.

Why does not the whole transmission line radiate in this way? The parallel conductors produce overlapping fields in antiphase which cancel out. (If you try to propagate on a line a frequency for which the separation of the conductors is a significant fraction of the wavelength the line *will* radiate!)

The energy radiates out from the dipole in all directions, though not with equal power density (in a direction along the dipole the power density is zero). The surfaces at right angles to the propagation are concentric spheres: these are equiphase surfaces for the electric field and for the magnetic field, and are often described as *wavefronts*. Just as described for the plane wave, the electric and magnetic fields are at right angles to each other, in phase and at right angles to the direction of propagation; the velocity of propagation (both phase and group velocity) equals $1/\sqrt{\mu_0 \cdot \varepsilon}$; the ratio E/H at a point equals $\sqrt{\mu_0/\varepsilon}$; the power density at a given point equals $E \times H$ (rms values).

Reduction of power density with distance in a spherical wave

In the absence of attenuation, a plane wave would show no reduction of power density with distance; if, however, the wavefronts are spherical, the power density falls off as the square of distance. This is because the same total power passes through spheres of increasing radius: the surface area of a sphere is proportional to its radius squared, so the power per unit area must be inversely proportional to the radius squared.

A spherical wave can be turned into a 'beam', that is an approximately plane wave but with a restricted area, by means of a lens or parabolic reflector. However, in free space, at a distance that is large compared to the initial beam diameter, the wavefront becomes effectively spherical again (an area on the surface of a sphere, not the whole sphere), due to the phenomenon of diffraction. In consequence, the power density of free-space radiation always falls off ultimately as the square of the distance from the driving point.

3.3 PLANE WAVE PROPAGATING BETWEEN TWO CONDUCTING PLANES

Figure 3.4 shows a cross-section of the fields associated with a wave on an ideal twin-wire transmission line – notice that here also the electric and magnetic fields are everywhere at right angles, and they are in phase. Now imagine that instead of two wire conductors we had two very wide flat conducting sheets; the equivalent wave to that of Fig. 3.4 would be as shown in Fig. 3.5 (ignoring what happens at the edges of the sheet). The wave shown in Fig. 3.5 is a plane wave, but the electric field is restricted in length to the distance between the sheets and terminates on charges in the sheets. The magnetic field only exists between the sheets, and in each sheet there are currents consistent with the magnetic field. *See* Fig. 3.6.

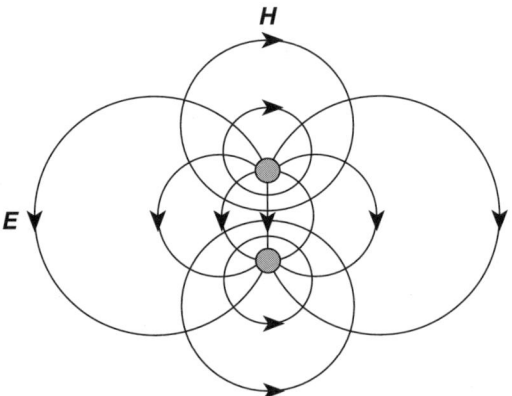

Fig. 3.4 Electric and magnetic fields associated with the wave on a twin-wire line (propagating into the page)

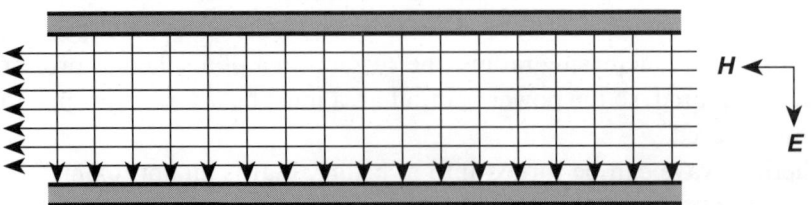

Fig. 3.5 A plane wave propagating between two conducting sheets (into the page)

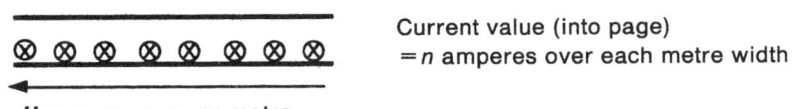

Current value (into page)
= *n* amperes over each metre width

H = *n* amperes per metre

Fig. 3.6   Relationship between current in a conducting sheet and magnetic field alongside it

3.4   REFLECTION AND REFRACTION OF PLANE WAVES

When an electromagnetic wave meets the surface of a good conductor it is reflected. When it meets the surface of a transparent material with a different dielectric constant from that in which it is propagating, generally part of it is reflected and part is transmitted. The transmitted part suffers a change of direction: this effect is called *refraction*. These phenomena are often studied in elementary physics, in the context of light, using the notion of *rays*; a ray may be equated with a directed line indicating the direction of wave propagation.

For simplicity in the discussion that follows, only plane surfaces will be considered.

Reflection by a good conductor

When an electromagnetic wave meets the boundary between two media its subsequent propagation is determined by what are known as the *boundary conditions*. It is known from experiment that a plane metal surface produces *regular reflection* – the angle to the surface of the reflected ray is equal to the angle of the incident ray – this can be shown to be consistent with the following boundary conditions.

At the surface of a conductor (assumed to have negligible resistivity):

- The resultant electric field cannot have a component along the surface – if such a field component tries to grow it causes redistribution of charge in the surface which cancels it.

   If a resultant electric field exists at the surface it must meet the surface at right angles and terminate, at the surface, on the appropriate charge.
- The resultant magnetic field cannot have a component normal (at right angles) to the surface – if such a field component tries to grow within the conductor it induces currents which generate an equal and opposite field (Lenz's law).

   If a resultant magnetic field exists at the surface it must be along the surface and be associated with the appropriate current in the surface.

When considering a transmission line terminated in a short circuit we used what was, in fact, a boundary condition – that the short circuit can have current in it but no voltage across it. This determined that the energy must be reflected back in a pulse with inverted voltage. The case now being considered is more

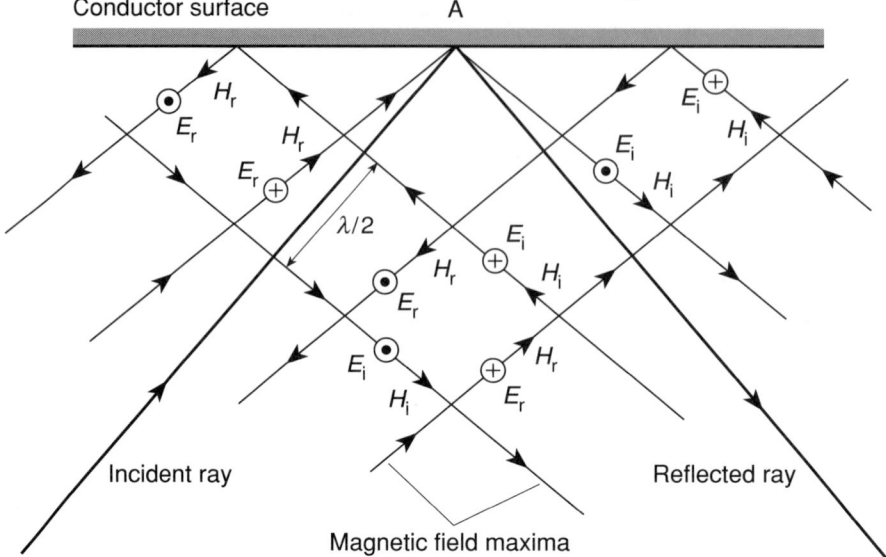

Fig. 3.7   Reflection by a plane conductor of a plane electromagnetic wave polarized parallel to the surface

complicated, being in three dimensions rather than one, but a similar outcome is to be expected; currents flow in the surface to generate a reflected wave the phase and direction of which are such as to satisfy the boundary conditions. It turns out that the direction of the reflected wave is always that indicated by the law of regular reflection – i.e. angle of incidence = angle of reflection.

Figure 3.7 shows the special case of a plane sinusoidal wave meeting a conductor surface having its electric fields along the surface. In the diagram the directions of propagation of the wave before and after reflection are indicated by an incident ray and a reflected ray. The fields are shown 'frozen' at an instant of time; even then it can only show magnetic field maxima, by lines on the page perpendicular to the rays, and electric field maxima, by a dot in a circle to indicate fields in the E–H plane which are out of the page and a cross in a circle for fields into the page. There are, of course, fields between the field maxima of sinusoidally differing amplitude. At the point A and, in fact, all along the surface, the electric field in the reflected wave cancels the electric field in the incident wave. As time progresses the incident wave moves towards the surface and the reflected wave away. If a point like A is considered, where field maxima in the incident and reflected waves coincide at the surface, it should be seen that the cross-over pattern of the fields moves to the right along the surface.

Refraction and total internal reflection

The case which is important for our purposes is that of a wave in a dielectric (medium 1) which meets a boundary beyond which is a material with a lower

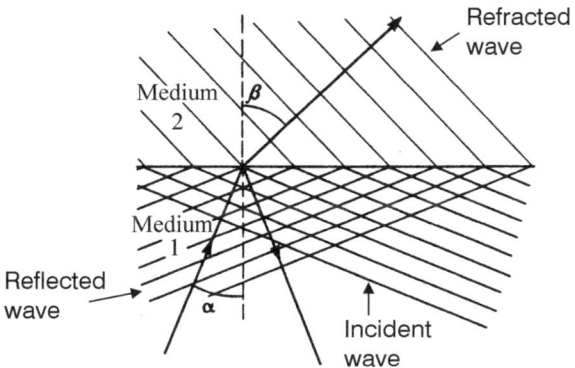

Fig. 3.8 Reflection and refraction of a plane electromagnetic wave at a dielectric boundary

relative permittivity (medium 2). Compared to medium 1, a sinusoidal wave in medium 2 has:

- A higher phase velocity ($1/\sqrt{\mu_0 \varepsilon}$) and therefore a longer wavelength (because its frequency does not change and its phase velocity is $f\lambda$).
- A greater wave impedance ($\sqrt{\mu_0/\varepsilon}$).

The boundary conditions at a dielectric boundary are not so straightforward as those at a conducting surface, so only their consequences will be described.

The effect depends upon the angle which the incident wave makes to the boundary. Figure 3.8 again takes the special case of a sinusoidal wave with electric fields parallel to the boundary (the results are broadly applicable to the more general case): only the magnetic field maxima are shown. With a small angle of incidence (between the ray and a normal to the surface) the refracted wave direction changes as shown to accommodate the increase in wavelength without a wavefront discontinuity at the boundary. The rule which gives the angles is known in optics as Snell's law, and is usually stated as

$$n_1 \sin \alpha = n_2 \sin \beta$$

where n_1 and n_2 are the *refractive indices* of media 1 and 2 respectively.[2] Snell's law could equally well be written

$$\frac{\sin \beta}{\sin \alpha} = \sqrt{\frac{\varepsilon_{r1}}{\varepsilon_{r2}}}$$

The reflected ray obeys the usual law of reflection. The transmission line analogy (but one-dimensional) would be a line connected to a second, correctly terminated, line of higher characteristic impedance. From this one can infer that the reflected wave suffers no electric field phase inversion. The ratio of intensities of

---

[2] The refractive index of a medium is the ratio of the velocity of light in free space to the velocity of light in the medium – i.e. $n = c/v_p$ – and so is directly proportional to the square root of the permittivity.

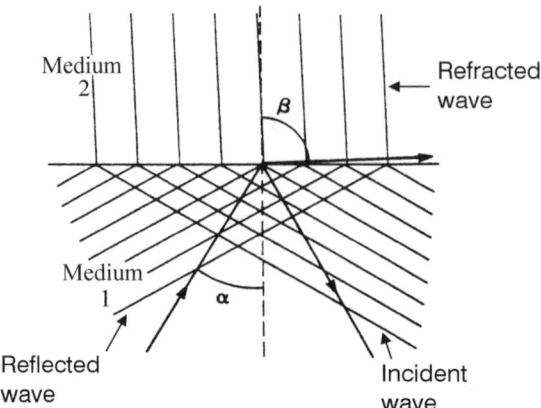

Fig. 3.9 Reflection and refraction at almost the critical angle

the transmitted and reflected waves depends upon the relative wave impedances of the two media and on the angle of incidence.

If the angle of incidence is increased the situation soon reaches that shown in Fig. 3.9 where the refracted ray is almost along the surface. As the angle of incidence increases further the condition in medium 2 changes: Snell's law cannot apply because $\sin \beta$ would have to be greater than unity – which is impossible. The value of the angle of incidence, α, for which Snell's law gives a value for $\sin \beta$ of unity is called the *critical angle*.

Fields still penetrate medium 2, but now their amplitude falls off exponentially with distance from the boundary, as shown in Figs. 3.10(a) and 3.10(b). Under these conditions the fields in medium 2 are described as *fringing* or *evanescent* fields. Since magnetic flux is continuous, the magnetic field in medium 2 has a longitudinal component as indicated. It can be seen that the distance between field maxima in the fringing fields is no longer the plane-wave wavelength in medium 2. The field patterns move to the right (for these diagrams) as time progresses. The moving field pattern in medium 2 is sometimes described as a *surface slow-wave*; the important point is that some energy is stored in these fields.

At the critical angle the reflected wave is still in phase with the incident wave. As the angle of incidence is increased there is an increasing phase shift of the electric field with reflection which, at grazing incidence, becomes inversion. The power density of the reflected wave is equal to that of the incident wave: the analogy with a transmission line is that the surface is now acting as a purely reactive load.

The condition just described is known as *total internal reflection* – a name which may give the false impression that no electromagnetic energy enters medium 2.

Comparison of conductor reflection and total internal reflection

When a wave is reflected at a conducting surface the electric field is inverted, whereas for total internal reflection at a dielectric boundary the electric field suffers a phase shift which depends on the angle of incidence.

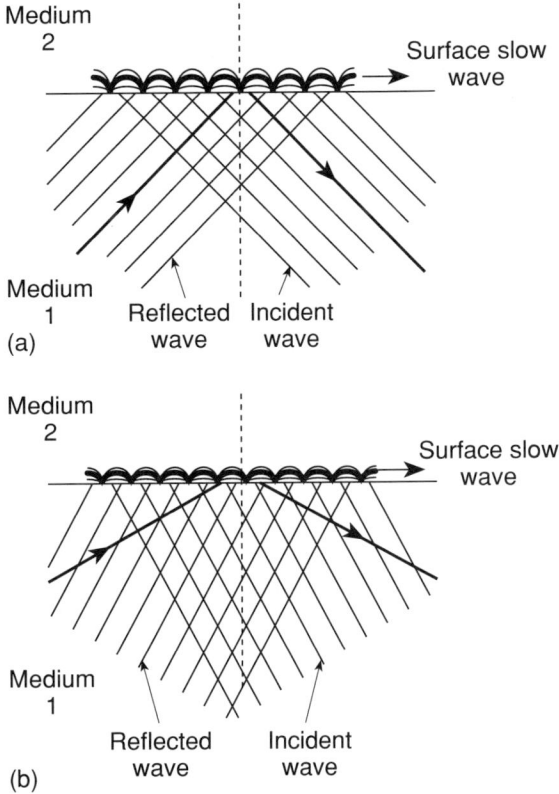

Fig. 3.10 The result of reflection of an electromagnetic plane wave on a dielectric boundary where the angle of incidence exceeds the critical angle

A reflecting conductor surface carries currents, which will cause losses if the conductivity is less than infinite. A reflecting dielectric carries fringing fields which will introduce losses if that dielectric is lossy. (There may also, of course, be losses in the propagating dielectric.)

For total internal reflection the angle of incidence of the wave must exceed the critical angle: the greater the angle of incidence (i.e. the smaller the angle which the ray makes to the surface) the less the extent of penetration of the fringing fields.

4 Coaxial line, strip line and waveguide

These three media can all be used at much higher frequencies than twin-wire line and their uses to some extent overlap, as we shall see.

4.1 COAXIAL LINE (COAX)

Coaxial line is used extensively to transmit signals in the frequency range 1 to 100 MHz over hundreds of metres. It is also used to transmit much higher frequency signals – up to 10 GHz – over short distances. The treatment given here is brief only because most coax properties are directly derivable from those of twin-wire line.

Coax consists of a wire conductor held in place along the centre of a conducting cylinder by a tube of dielectric. The outer conducting cylinder is usually made from woven copper wire so as to give the line a degree of flexibility. For a given dielectric, the relative radii of the inner and outer conductor determine the characteristic impedance (*see* Appendix 2). For low loss and high power rating the overall cross-sectional dimensions need to be as large as possible, but larger dimensions lead to less mechanical flexibility and also allow higher order modes to propagate at lower frequencies (*see* later section on waveguide).

The signal is applied between the inner and outer conductors of the line and the circuit analysis given for twin-wire line applies, although the inner conductor has most of the line resistance. Coax tends to be more dispersive than twin line at low frequencies, but its usefulness extends to much higher frequencies than twin line because the fields are mainly restricted to the inside of the outer conductor resulting in much less radiation and cross-talk. A great deal of coax is made to one of two standard characteristic impedances; either $75\,\Omega$ or $50\,\Omega$.

A cross-section of the fields in a coaxial line in the normal mode of propagation is shown in Fig. 4.1. The wave is a sort of plane wave wrapped round the centre conductor; it does not have a direction of polarization. The velocity of propagation of a high-frequency signal in coax is that of an electromagnetic plane wave in the dielectric between the conductors; typically 2×10^8 m/s.

Fig. 4.1   Electric and magnetic fields associated with the propagation of a wave in coax (into the page)

Coaxial cable

A coaxial line encased in an insulating protective sleeve is usually referred to as a coaxial cable. This is an exception to the normal usage in which a cable consists of an assembly of a number of lines carried together in the same protective sheath.

Balance about earth

Usually the outer conductor of a coaxial line is connected to earth potential at least at one end. This is in contrast to a twin-wire line which in normal use is balanced about earth. If a length of coax is used to feed a load, such as an antenna, which needs a balanced feed, then a special 'balance to unbalance' transformer has to be included (*see* Chapter 7).

4.2   STRIP LINE

Strip line is usually encountered in the form of microstrip, used to convey signals from one part of a high-frequency circuit to another. When a circuit has to handle frequencies such that the distance between components is significant in wavelengths the connections must be formed as transmission lines which can be matched to the input and output impedances of the components. Often, different specified characteristic impedances are required for different parts of a circuit. The microstrip is formed on printed circuit board with a ground plane acting as one conductor, as shown in Fig. 4.2.

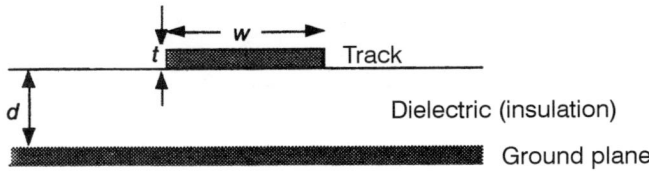

Fig. 4.2   Microstrip

The velocity of signal propagation in microstrip depends solely on the permittivity of the dielectric. The characteristic impedance depends on the thickness of the dielectric, d, the width of the track, w, and to some extent on the thickness of the track, t. In general, increasing d increases Z_0 while increasing w decreases it. Characteristic impedances in the range $50\,\Omega$ to $150\,\Omega$ are easily achieved. Sometimes other conductors close-by will modify Z_0. Generally the required dimensions of microstrip in a circuit to provide the correct impedances will have to be worked out using computer aided design. For further comments see Appendix 2.

4.3 WAVEGUIDE

Waveguide is used for moving signals of very high frequencies over short distances – commonly for connecting a microwave antenna to its transmitter and receiver. Propagation in optical fibre is much easier to explain when propagation in metal waveguide is understood, so for this reason, if for no other, we need to discuss it in some detail.

Reflection between two parallel conducting planes

A plane wave can propagate between two parallel conducting planes by multiple reflection. Consider the simplest case where the direction of polarization of the wave (that is, the electric field direction) is parallel to the planes.

The reflections from the two planes have to be at such an angle that they reinforce rather than destructively interfere. To show what this implies, Fig. 3.7 has been redrawn and extended in Fig. 4.3 (leaving out the electric field) and a possible position added in which a second conducting plane could be placed so that the wave and its reflection in the first plane together satisfy the boundary conditions

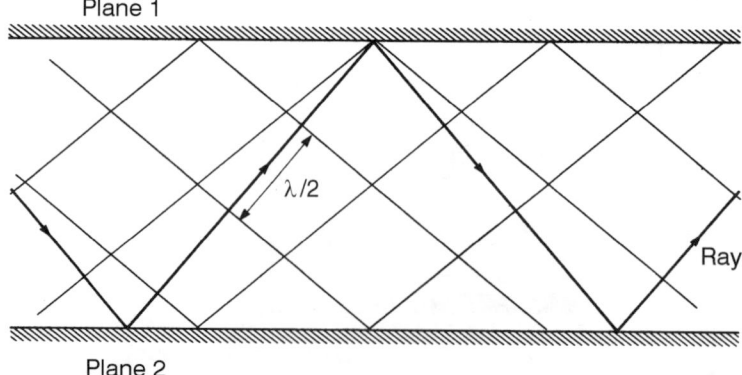

Fig. 4.3   Reflection of a plane electromagnetic wave between two conducting planes

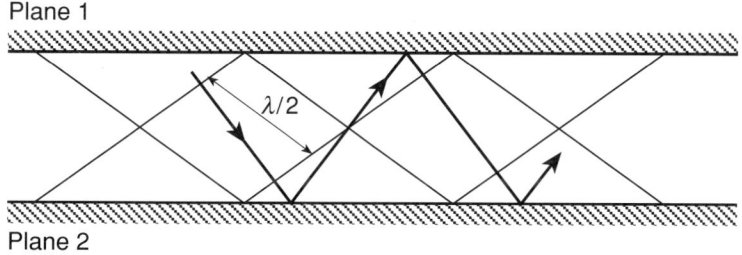

Plane 1

$\lambda/2$

Plane 2

Fig. 4.4   Reflection of a plane electromagnetic wave between two planes with the minimum separation for the wavelength

at this second plane. If a second plane were in fact placed in this position, then the wave would be entirely contained between the two planes and propagate to the right in the diagram by multiple reflection. The magnetic field maxima in the overlapping waves produce a 'diamond pattern': the second reflecting plane could have been placed so as to include just one diamond width, as shown in Fig. 4.4, or three, or any number.

Rectangular waveguide

Two more conducting planes could be added to the two described in the previous paragraph, at right angles to the electric field, so as to terminate the electric field on charges in the conductors. These conductors would carry currents consistent with the magnetic fields tangential to them (as indeed do the reflecting planes). The resulting conducting tube, of rectangular cross-section, is a *waveguide*. For the moment, consider the simplest possible propagation down this tube with one diamond pattern across between the reflecting walls. Figure 4.5 shows the magnetic field maxima of the reflecting wave at a particular instant of time, together with its direction of propagation, in a plane perpendicular to the reflecting walls. The energy is carried by the plane wave, which will be referred to from now on as the generating wave, and so it progresses along the waveguide with a group velocity

$$v_{g(\text{guide})} = v_p \cos \theta$$

where v_p is the phase velocity of the generating wave (c, if the waveguide contains air or a vacuum). Suppose that in the same waveguide one tried to propagate a

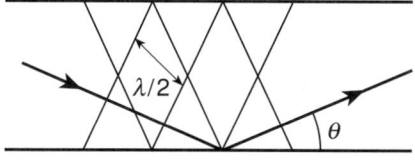

$\lambda/2$

θ

Fig. 4.5   Direction of propagation of the generating (reflecting) plane wave in a rectangular waveguide

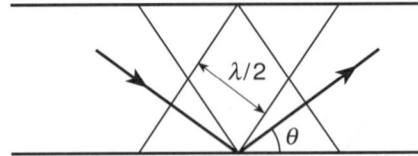

Fig. 4.6   Direction of propagation of the generating wave in a waveguide for a longer wavelength than in Fig. 4.5

lower frequency. The free-space wavelength, λ, is longer. The direction of reflection would have to adjust itself as shown in Fig. 4.6. The angle θ is increased. The group velocity is reduced. A group velocity which varies with frequency indicates that a waveguide propagating in this way is dispersive and the dispersion that results is called *waveguide dispersion*.

Suppose one continues to reduce the frequency of the wave one tries to propagate. When the half-wavelength becomes equal to the distance between the reflecting walls the generating wave will reflect between the walls making no progress at all. This frequency is called the *cut-off frequency*: below this frequency, wave propagation in the guide is impossible.

Modes

Look back at Fig. 4.5. A wave of the wavelength shown could also propagate as shown in Fig. 4.7: this is known as a different *mode* of propagation. It will be found if an attempt is made to construct the diagram that the wave shown in Fig. 4.6 cannot propagate in this mode because its wavelength is too long. Looking at Figs. 4.5 and 4.7 it should be clear that energy in the two modes propagates with very different group velocities. There are several ways in which energy can be launched into a waveguide; one is discussed at the end of this chapter. Whatever method is used, if both the modes shown in Figs. 4.5 and 4.7 can propagate then there will be energy in each mode. Even if the launching section is carefully designed to encourage one mode rather than another, small irregularities in the guide structure will cause energy to be transferred from one mode to the other as the waves progress. The difference in group velocity between the two modes results in a form of dispersion known as *intermodal dispersion*.

In fact there are many modes that can propagate. The generating wave could reflect from the other pair of walls, or indeed from all four walls in the most

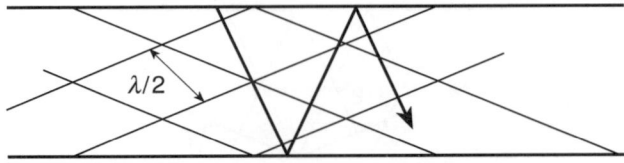

Fig. 4.7   Propagation of a wave of the same wavelength as that in Fig. 4.5 in a higher mode

general case. Each of the possible modes has its own cut-off frequency; the mode described first has the lowest and is called the *fundamental* mode. Provided the cross-section of the guide has a two to one aspect ratio there is a two to one ratio of frequency over which only the fundamental mode can propagate. If frequencies in this single-mode range are propagated there can be no intermodal dispersion.

Composite field patterns

As the reflecting generating plane wave crosses and recrosses itself the resulting composite field pattern – which is a form of interference pattern – looks nothing like a plane wave. In *transverse electric* modes the magnetic fields add vectorially to form loops while the electric fields add in some places and cancel in others. The resulting pattern for the fundamental mode is shown in Fig. 4.8. This pattern travels down the guide and its speed of travel, the phase velocity in the guide, is greater than the plane wave phase velocity. To understand why this is so imagine a closing pair of scissors: the point of intersection of the blades (equivalent to the field pattern) moves much faster than the speed at which the blades (equivalent to the reflecting generating wave) move towards each other. Because the field pattern travels faster than the energy, as a pulse of electromagnetic energy travels along the guide the field pattern travels through the pulse, taking on a finite magnitude at the beginning of the pulse and collapsing to zero at the end of it.

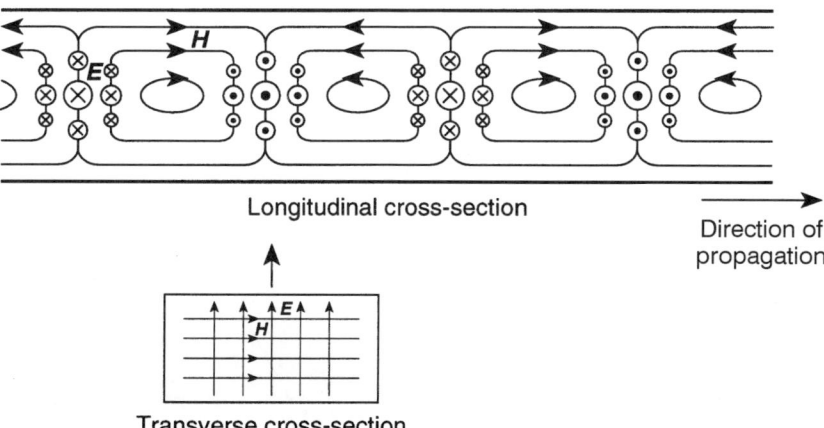

Longitudinal cross-section

Direction of propagation

Transverse cross-section

Fig. 4.8 Composite field pattern for the fundamental mode in a rectangular waveguide

Each mode is a solution to Maxwell's equations in its own right; in fact any distribution of fields that satisfies Maxwell's equations and the boundary conditions at the waveguide walls will propagate as a mode. Since, however, in every case for rectangular guide, a generating plane wave can be postulated, an approach from that point of view is found to be the more useful.

Analysis of the fundamental mode in rectangular guide

In practical use, propagation in rectangular guide is always in the fundamental mode, so from now on we shall concentrate on that. Look at Fig. 4.9 which shows just one magnetic loop of the composite field pattern, together with half wavelengths of the generating wave. The mode pattern will have a wavelength, designated λ_g, which is the distance along the guide at any instant between two adjacent points with the same electric and magnetic field phases: by inspecting Fig. 4.8 you should see that one magnetic loop is half a guide wavelength long, and this is marked on Fig. 4.9. θ is the angle made by the direction of propagation of the generating wave with the axis of the guide: simple geometry indicates the equality of the two angles marked θ. For the phase velocity of the mode we can write

$$v_{p(guide)} = f\lambda_g$$

so, since, assuming that the waveguide contains no dielectric other than air

$$c = f\lambda \qquad v_{p(guide)} = \frac{c\lambda_g}{\lambda} \tag{4.1}$$

Looking at Fig. 4.9 it can be seen that as the generating wave moves forward through $\lambda/2$ the field pattern moves forward through $\lambda_g/2$, and λ_g is always greater than λ. Hence $v_{p(guide)}$ is always greater than c.

It has already been seen that

$$v_{g(guide)} = c\cos\theta$$

and from Fig. 4.9

$$\cos\theta = \frac{\lambda/4}{\lambda_g/4} = \frac{\lambda}{\lambda_g}$$

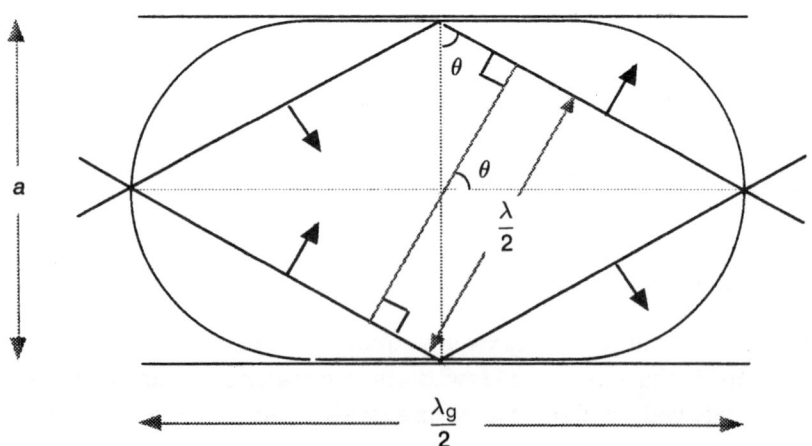

Fig. 4.9   One magnetic loop of the fundamental mode

so

$$v_{\text{g(guide)}} = \frac{\lambda}{\lambda_{\text{g}}} \tag{4.2}$$

From Equations (4.1) and (4.2)

$$v_{\text{p(guide)}} \times v_{\text{g(guide)}} = c^2$$

Guide wavelength formula

A relationship between λ_{g} and λ for the fundamental mode can be deduced as follows. From Fig. 4.9

$$\sin \theta = \frac{\lambda/4}{a/2} = \frac{\lambda}{2a}$$

Since $\cos \theta = \lambda/\lambda_{\text{g}}$ and $\sin^2 \theta + \cos^2 \theta = 1$

$$\frac{\lambda^2}{(2a)^2} + \frac{\lambda^2}{\lambda_{\text{g}}^2} = 1 \tag{4.3}$$

At the cut-off frequency the wavelength is twice the waveguide width, i.e. we can write

$$2a = \lambda_{\text{c}}$$

which, in Equation (4.3) gives

$$\frac{1}{\lambda_{\text{g}}^2} = \frac{1}{\lambda^2} - \frac{1}{\lambda_{\text{c}}^2} \tag{4.4}$$

Taking certain assumptions from the graphical analysis above, Appendix 4 gives an algebraic analysis of the fundamental mode in rectangular guide, which reveals a further interesting phenomenon; that of evanescent propagation.

Guide wave impedance

At every point in a travelling wave in the fundamental mode in rectangular waveguide, the ratio of the transverse electric and magnetic fields has a fixed ratio: this is called the guide wave impedance, Z_{W}. (Notice that the longitudinal component of the magnetic field is not involved in the definition.) In Appendix 4 it is shown that

$$Z_{\text{W}} = 120\pi \frac{\lambda_{\text{g}}}{\lambda}$$

This wave impedance is analogous to the characteristic impedance of a transmission line: a resistive film of surface resistivity equal to Z_{W} placed across the end of a guide carrying the fundamental mode would absorb all the energy and match the guide.

Waveguide wall currents and voltages

As already indicated, the waveguide walls carry currents and voltages consistent with the fields inside the guide. In fact, a complete description of the propagation could be given in terms of these currents and voltages – this does not, however, prove to be a very productive approach. A general illustration of the configuration of the wall currents in the fundamental mode is shown in Fig. 4.10; voltage maxima occur where the current causes charge to accumulate.

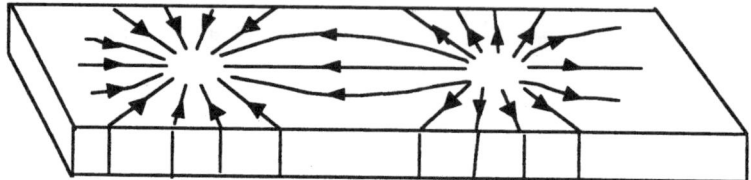

Fig. 4.10   Currents in the walls of a waveguide carrying the fundamental mode

Attenuation

Since waveguide runs are normally short, attenuation in practical use is not normally a problem, and in the discussion so far it has been ignored. Unless the waveguide contains a solid dielectric (which will dissipate energy as the wave travels through it) losses in a waveguide are caused by the currents in the walls. These currents flow in a very thin section of the inner surface (the *skin effect*), so to minimize attenuation the inside surface of the guide needs to be clean – particularly the corners, where dirt tends to accumulate. Very high quality waveguide is made from very pure copper, and is even sometimes silver plated inside: cheaper waveguide is made of brass.

In theory, the attenuation due to wall currents is infinite at the cut-off frequency and becomes small rapidly as one moves to higher frequencies. The group velocity also changes rapidly near the cut-off frequency, giving high dispersion. For both these reasons the range of frequencies recommended for a given size of waveguide extends from about 25 per cent above the cut-off frequency of the fundamental mode to about 90 per cent above. The latter restriction is to avoid the possibility of generating a higher mode.

β_g/ω curves for rectangular waveguide

Figure 4.11 shows the shape of β_g/ω curves for the fundamental mode and for two nearest competing modes in a rectangular waveguide with a 2:1 aspect ratio, β_g being the phase change coefficient for the mode in the guide. (The names of modes in rectangular waveguide tell you which field is purely transverse and how many half-cycles of field variation there are across each guide dimension.) At C_{10}, the cut-off angular frequency of the fundamental mode (TE_{10}), the

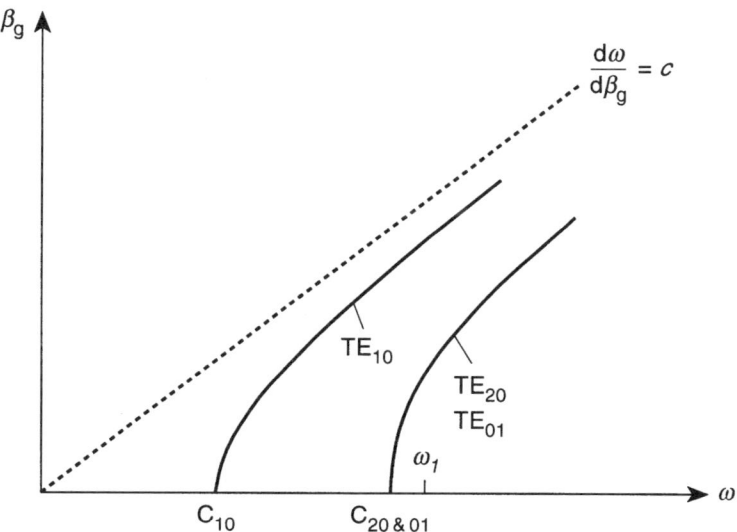

Fig. 4.11 β_g/ω curves for lower order modes in rectangular waveguide

slope, $\mathrm{d}\beta_g/\mathrm{d}\omega$, is infinite, so $v_{\mathrm{g(guide)}}$, i.e. $\mathrm{d}\omega/\mathrm{d}\beta_g$, is zero. At this point, ω/β_g, the value of $v_{\mathrm{p(guide)}}$, is infinite. As the frequency increases, the inverse slope, representing $v_{\mathrm{g(guide)}}$, increases towards a value c, while, since the curve is asymptotic to the line of slope $1/c$, the value of $v_{\mathrm{p(guide)}}$ decreases towards a value c. The frequency range between C_{10} and $C_{20\&01}$ is the single-mode range of operation of the guide. The fact that the β_g/ω lines are curved indicates waveguide dispersion, while it should be possible to see that at the angular frequency marked ω_1, a frequency where more than one mode can propagate, the group velocity in the fundamental and in the higher modes is different, indicating intermodal dispersion.

Circular waveguide

Waveguide with a circular cross-section (or indeed almost any cross-section) can propagate waves. Again, any field pattern that satisfies Maxwell's equations and the boundary conditions at the wall will propagate. The relationship between phase and group velocities is as described for rectangular guide.

Circular guide also has a fundamental mode which has a similar longitudinal field distribution to the fundamental mode in rectangular guide – *see* Fig. 4.12 for the transverse field cross-section. Circular guide has the disadvantage that the structure does not constrain the direction of polarization for this fundamental mode.

A coaxial cable can be thought of as a circular waveguide with an extra conductor down the middle, and hence it is possible for modes other than the simple mode described at the beginning of this chapter to propagate. This is not usually a problem, since the small diameter of the outer conductor means

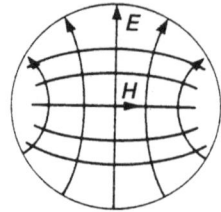

Fig. 4.12 Transverse field cross-section in the fundamental mode in circular waveguide

that generally all alternative modes are cut off, but at high enough frequencies they might not be.

Termination of a waveguide

A waveguide in use needs to be correctly terminated just as does a line, to avoid reflections. A termination with a uniform surface resistivity equal to the wave impedance is not generally a practical proposition. Attenuating material placed in the guide with an increasing volume over a number of wavelengths can be designed to absorb the wave without reflection, but this simply turns the wave energy into heat.

The signal can be coupled into the waveguide from a coaxial line using the inner conductor of the line as a sort of antenna protruding into the guide through the broad wall as shown in Fig. 4.13. The distance to the short-circuiting end wall at A is nominally $\lambda_g/4$ so that the wave travelling towards the wall, after reflection in which it suffers a phase reversal, travels back in phase with the wave travelling from the probe directly down the guide. There are a number of variants on this theme, and the detail of design is complex, generally requiring CAD (computer aided design). It is also possible for the signal to be coupled in from a resonant cavity using a quarter-wave transformer (*see* Chapter 7). The signal can be coupled out at the receiving end using the same structures.

The concepts of reflection coefficient, return loss and standing wave ratio are relevant to waveguide.

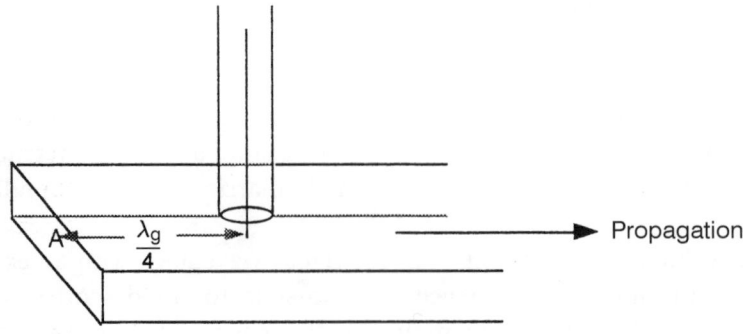

Fig. 4.13 Wave-launching section in a waveguide

4.4 WAVEGUIDE CALCULATIONS

4.1 Signals from a UHF television transmitter to its antenna are carried by waveguide. Calculate suitable dimensions for rectangular waveguide to carry signals in Band V, which covers a frequency range 614 to 854 MHz.

4.2 Calculate the guide wavelength, phase velocity, group velocity and wave impedance of a 6.0 GHz travelling wave in a waveguide of internal dimensions 4.0 cm by 2.0 cm.

5 Optical fibre

Optical fibre is a type of dielectric waveguide used to propagate radiation in the infra-red to visible region of the spectrum.

By convention, optical frequencies are not generally quoted directly. This is because it is not easy to measure them directly; instead free-space wavelength is measured and quoted, usually, in nanometres. Mention of 'wavelength' in what follows should be taken to mean free-space wavelength; when the wavelength in the fibre needs to be specified it will be referred to specifically as *guide wavelength*. In place of the term *bandwidth* representing the range of frequencies to be transmitted, reference is made to *linewidth* representing the spread of (free-space) wavelengths: it is easy to show that the linewidth as a percentage of the centre wavelength is the same as the bandwidth as a percentage of the centre frequency.

In all practical systems up to the time of writing the transmission is binary digital; electrical pulses representing 1s are converted into bursts of light and launched onto the fibre using light-emitting diodes (LEDs) or lasers and received and converted back into electrical pulses using photodiodes. An elementary explanation of the function of these devices is given in *Introduction to Solid State Devices*, Lem Ibbotson, Arnold, 1997. For the present discussion you need to know that none of the transmitting devices, even when operating continuously, emits a single frequency of radiation: the output of an LED has a linewidth, typically, of about 60 nm, a laser has a linewidth of 5 to 10 nm, while a 'single mode laser' can have a linewidth of less than 0.01 nm. There is also, of course, a spread of frequencies associated with the pulse modulation which would occur even if the source were monochromatic (single frequency), however this is only significant compared to the narrowest source linewidths at the very highest signalling rates.

High quality fibres are generally made of amorphous silica (silicon dioxide) with small quantities of suitable doping material added to alter the refractive index as required. There is available some cheap fibre made of plastic, but this is only suitable for transmission over very short distances, and we shall not discuss it further.

The fibre consists of a cylindrical core surrounded by a cladding of lower refractive index. It may be:

- Single mode (step-index) with a core diameter, for the wavelengths used in telecommunications $\approx 9\,\mu m$, or

- Multimode step index with a core diameter $\approx 50\,\mu m$ or
- Multimode graded index with a core diameter $\approx 50\,\mu m$ and a core refractive index that varies across the core diameter, being maximum at the centre.

The cladding outer diameter is typically $\approx 125\,\mu m$.

5.1 PROPAGATION IN OPTICAL FIBRE

Propagation can be modelled as a generating wave suffering repeated reflections by *total internal reflection* at the core boundary; however, since no charge can accumulate at the boundary, nor current flow in it, the nature of the reflection differs fundamentally from that at a metal boundary. The fields penetrate into the cladding to an extent that depends on the frequency and on the angle of incidence of the generating wave to the boundary, and these *evanescent fields* travel along in the cladding in step with the composite fields in the core and with an intensity that falls off exponentially from the boundary. The cladding is designed to be wide enough effectively to contain these fields under all circumstances. Because of the varying extent of the field penetration into the cladding, the fibre is said to have a *soft boundary*.

Many modes are possible. The propagation properties of the fibre can be illustrated by drawing a β_g/ω diagram as for rectangular metal waveguide. In the case of fibre, however, it proves more useful to plot 'normalized' values, β_g/β_0 and ω/ω_0. The normalizing parameter β_0 is the 'free-space' propagation constant for that frequency, i.e. $2\pi/\lambda$, while ω_0 is a constant that emerges from the detailed theory (*see*, for example, John Gower, *Optical Communication Systems*, 2nd Edition, Prentice Hall).

$$\omega_0 = \frac{c}{a\sqrt{n_1^2 - n_2^2}}$$

so that

$$\frac{\omega}{\omega_0} = \frac{2\pi a}{\lambda}\sqrt{n_1^2 - n_2^2}$$

where a is the core radius and n_1 and n_2 are the refractive indices of core and cladding respectively. [ω/ω_0 is sometimes given the symbol V in the literature.]

Figure 5.1 shows theoretical normalized β_g/ω curves for some of the lower order modes in step index silica fibre, assuming that the refractive indices of core and cladding are independent of frequency (the significance of this proviso will become clear later). The names given to the modes represent the composite field distributions in a scheme which no attempt will be made to explain here.

Since refractive index is the ratio of phase velocity in free space to that in the medium it is also the ratio of phase change coefficient in the medium to that in free space (i.e. β_g/β_0). It can be seen that on the β_g/β_0 axis, the values of n_1

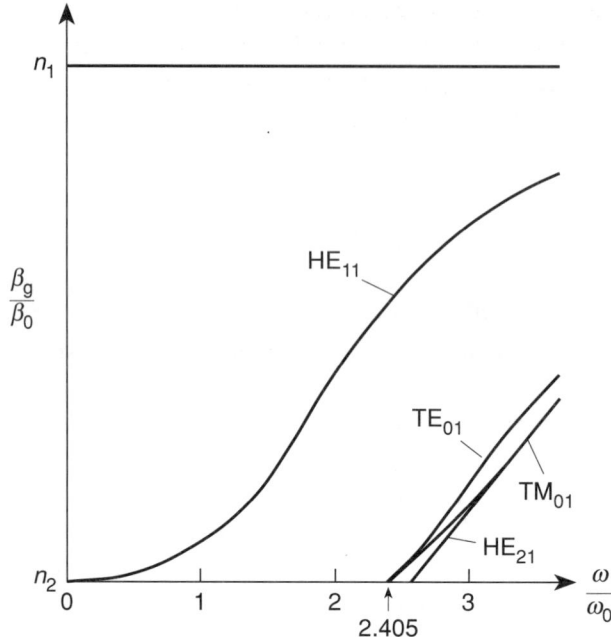

Fig. 5.1 Theoretical normalized β_g/ω curves for step-index silica fibre

and n_2 are marked: the average refractive index encountered by the wave moves nearer to n_2 as the wavelength increases, implying that an increasing proportion of the wave is propagating in the cladding. In fibre a mode cuts off when the value of β_g/β_0 reaches n_2, indicating that the core is no longer guiding the mode.

It can be seen from Fig. 5.1 that there is a fundamental mode, called HE_{11}, which over a range of values of ω/ω_0 is the only mode that can propagate. The normalized β_g/ω curve for HE_{11} goes down to zero frequency, so that this mode has no lower cut-off frequency (the reason for this is that the soft boundary keeps expanding as the frequency is lowered); however, a minimum operating frequency is imposed by the attenuation properties of the fibre as will be discussed in the next section. Since the normalized β_g/ω line for HE_{11} is curved, there will be waveguide dispersion in this mode. The composite fields in the HE_{11} mode have some similarity to the fields in the fundamental mode in circular waveguide, but the electric field, as well as the magnetic field, has longitudinal components in the cladding.

By considering the slopes of the normalized β_g/ω curves for the different modes shown, at frequencies at which several can propagate, it can be seen that there will be significant intermodal dispersion.

5.2 ATTENUATION IN OPTICAL FIBRE

Glasses used for normal optical purposes absorb light to an extent which, though not noticeable in a lens or prism, would be far too great over a length of many kilometres. Transmission by silica optical fibre became a practical proposition when means were found of producing material with sufficient purity and homogeneity to give attenuation less than 20 dB/km. This can be achieved for wavelengths in the range 0.5 to 1.8 μm; in fact, at the optimum wavelength the attenuation can be as low as 0.2 dB/km.

There are two significant unavoidable loss processes in silica, both of which are frequency dependent:

- The silicon and oxygen inter-atomic links have resonances which absorb energy in the infra-red.
- The amorphous nature of the material leads to random fluctuations in density which cause a type of scattering, known as Rayleigh scattering; this decreases with increasing wavelength. The scattered energy leaves the fibre.

Figure 5.2 shows a graph of attenuation versus wavelength for a silica fibre. The dotted lines represent the effects of the two mechanisms described above; the solid line shows what is achieved in practice. The bumps in the practical curve are due to residual impurities which prove a practical impossibility to eliminate – particularly the large effect at 1.4 μm caused by water.

It can be seen that there are low-loss 'windows' at 1.55 μm and at 1.3 μm; long distance transmissions generally make use of one or other of these. Wavelengths in the small minimum around 0.85 μm are also used, mainly because cheap light sources and detectors are available for these wavelengths.

There are other glass-like materials in which the absorption edge is at a longer wavelength than that for silica, thus, since the effect of Rayleigh scattering reduces with increasing wavelength, use of such a material should result in lower

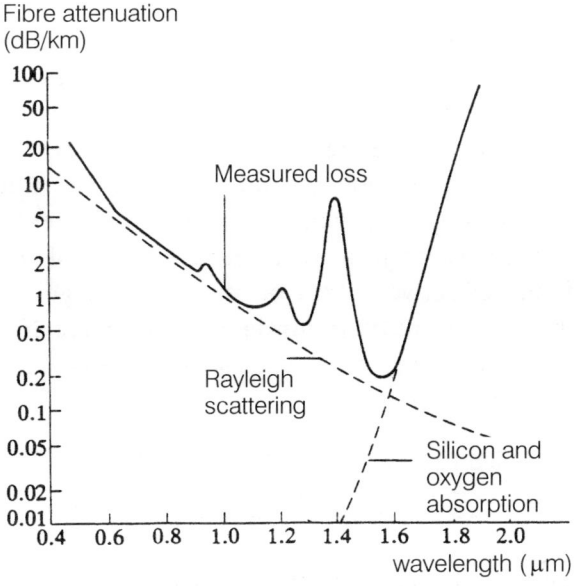

Fig. 5.2   Wavelength dependence of loss in silica fibre

attenuation still. Up to now these materials have proved too expensive and diffi-cult to process to be a practical proposition.

[An attenuation effect due to the semiconductor energy gap in silica has not been mentioned: this would only be significant for radiation at the violet and ultraviolet end of the spectrum.]

5.3   DISPERSION IN OPTICAL FIBRE

As in metal waveguide, both waveguide dispersion and intermodal dispersion can occur in optical fibre, but in addition there is another dispersive mechanism – *material dispersion*. This occurs because the phase velocity of plane waves in silica falls with increasing frequency. The effect of this in optical fibre is compli-cated because a fall in phase velocity implies a rise in refractive index, so that material dispersion indicates not only a change in the speed of the generating wave with frequency, but also a change in the effective diameter of the boundary.

In fibre designed to propagate many modes – multimode fibre – intermodal dis-persion is many orders of magnitude larger than dispersion due to the waveguide effect and that due to the material, so these last two can be ignored.

In single mode fibre the two remaining dispersive effects are significant, and they are found to interact. Thinking back to the explanation of propagation in metal waveguide it will be realized that a fall in the phase velocity of the generat-ing wave causes a fall in the mode group velocity, so material dispersion tends to reduce the group velocity as frequency increases. However, waveguide dispersion itself tends to increase the mode group velocity as frequency increases. (The detail is complicated because of the 'soft boundary' effect.) The change of phase velocity with frequency caused by the material dispersion is more rapid at shorter wave-lengths, giving over this range of wavelengths a net reduction in the group velocity with increasing frequency, whereas at longer wavelengths, as the effect of material dispersion decreases, the effect of waveguide dispersion increases and the group velocity increases with increasing frequency. Between these two regions is a wavelength at which the group velocity is not changing with frequency, so that monochromatic radiation would suffer no dispersion. The exact value of this wavelength can be manipulated to some extent by choice of the core diameter and core and cladding refractive indices because the waveguide dispersion will be different over different regions of the normalized β/ω curve of the fundamental mode (*see* Fig. 5.1).

5.4   LONG DISTANCE SINGLE MODE PROPAGATION AT 1.3 μm

To propagate signals successfully over many kilometres it is necessary to keep both attenuation and dispersion to a minimum. It has proved easy to design high quality step index silica fibre with a dispersion minimum at around 1.3 µm wavelength.

The fibre

Referring back to Fig. 5.1, it can be seen that in order that only the H_{11} mode can propagate, w/w_0 must be less than 2.405 (this rather odd theoretical value is the lowest value of w/w_0 for which the zero-order Bessel function is zero and arises because the fibre is cylindrical). We can write

$$\frac{2\pi a}{\lambda}\sqrt{n_1^2 - n_2^2} < 2.405$$

so, the core diameter must be

$$d < \frac{2.405\lambda}{\pi(n_1^2 - n_2^2)^{1/2}} \tag{5.1}$$

For ease of launching the signal into the fibre core and for reasons concerning mechanical tolerances in manufacture the core diameter needs to be as large as possible and this indicates that the difference between core and cladding refractive indices must be as small as practicable. However, it is sometimes necessary to bend a fibre to a moderate extent, and if the difference in refractive index is too small one finds that when the fibre is bent it leaks radiation. Refractive index values typically used are $n_1 = 1.460$ and $n_2 = 1.458$. Putting these values and $\lambda = 1.3\,\mu m$ into the formula gives the critical core diameter

$$d_c = 13\,\mu m$$

The standard value of core diameter used, 9 µm, is comfortably below this.

Example calculation
What is the shortest wavelength that can propagate in fibre with a core diameter of 9 µm and refractive indices of core and cladding of 1.460 and 1.458 respectively in the fundamental mode only?
   Transposing Equation (5.1)

$$\lambda_c = \frac{d\pi(n_1^2 - n_2^2)^{1/2}}{2.405}$$

$$= \frac{9 \times 10^{-6} \times \pi \times \sqrt{1.460^2 - 1.458^2}}{2.405}$$

$$= 0.9\,\mu m$$

Notice that radiation of wavelength 8.5 µm does not propagate as single mode in this fibre.

The radiation source

There is only one wavelength, for a given fibre, at which the dispersion is zero, and so, since all sources have some linewidth, there will be some dispersion. It is certainly necessary to use a laser rather than the much wider linewidth LED. The amount of dispersion (measured as the spread of a very narrow pulse) at the receiver depends on the linewidth of the source and on the distance that the pulse has travelled and so, for a given fibre, it is quoted as a number of pico-seconds per nanometre per kilometre.

Example calculation

For a given fibre at a signal wavelength of 1.3 μm the dispersion is 2 ps/nm per km (a typical value). What is the amount of dispersion at the receiver of a link of length 50 km if the linewidth of the source is 10 nm?

The answer is $2 \times 10 \times 50\,\text{ps} = 1\,\text{ns}$

Inevitably, the intensity profile of a light pulse from the source is not square; the oscillations need time to build up at the beginning of the pulse and to fall at the end, so that the pulse is shaped somewhat as shown in Fig. 5.3. This is no bad thing, since it reduces the effect of dispersion relative to a truly square pulse. It is normally assumed that for pulses like those shown in Fig. 5.3 the maximum amount of dispersion acceptable is half a symbol interval (the implication is that the effect of this dispersion is to make the trailing edge of the pulse reach the centre of the following symbol interval). Hence, the dispersion calculated in the example indicates a minimum symbol interval of 2 ns, and thus a maximum signalling rate of 500 Mbaud. To use higher signalling rates than this a source with a narrower linewidth would be required.

Fig. 5.3   Practical intensity profile of a typical pulse in optical fibre

Attenuation

The actual fibre attenuation at this wavelength is typically about 0.5 dB/km so that a 50 km span of continuous fibre would have an attenuation of 25 dB. The maximum acceptable attenuation depends on the maximum power available at the source and the minimum power required to work the receiver.

The maximum transmitted power depends on available sources and also on certain safety considerations. A typical figure for transmitter power is $\frac{1}{2}$ mW, i.e. −3 dBm. The minimum required receiver power is usually known as the *receiver sensitivity*: a typical value is −40 dBm. The difference between these two figures, 37 dB, suggests that there is plenty of power for a 50 km span, however this may not be the case, as will now be shown.

When fibre is manufactured it has to be coiled onto drums for transport to where it is to be installed; the greatest practical length of continuous fibre is limited by this to 2 km. Hence a 50 km span, say, has to be assembled by fusing together 25 lengths of fibre, requiring 24 fused joints or *splices*. Clever techniques are used for aligning the sections of fibre at the joints, but nevertheless, because of manufacturing and alignment tolerances, each joint presents a discontinuity and sets up a reflection. The reduction of transmitted power at each splice has to be estimated: it is usually taken to be 0.5 dB.

At each end of the link the fibre is connected to the transmitter and receiver by mechanical couplers which can be detached and reconnected. There must be at least two of these, one at each end, and they each account for an estimated 1 dB of power loss.

Finally, in planning a link, account has to be taken of the probability that repairs, entailing extra splices, will be necessary over the life of the link.

All this is usually presented in the form of a *link power budget*, as shown in Table 5.1.

Table 5.1   Power budget for an optical fibre link at a wavelength of 1.3 μm

transmitter power	−3 dBm
receiver sensitivity	−40 dBm
System margin	37 dB
fibre loss	25 dB
splice loss	12 dB
connector loss	2 dB
repair margin (10 splices)	5 dB
Route losses	44 dB
Excess margin	−7 dB

For a span to be usable the excess margin must be positive; hence a span of 50 km is not possible.

Of course, the system might work when first installed, before any repairs, and if the splices were made very carefully to keep the total splice losses below 10 dB, but it would be wrong to design it on that basis. A 40 km span would be possible.

Noise

The receiver sensitivity is decided by the minimum signal to noise ratio required at the output of the receiver where the decision in each symbol interval as to whether a pulse is present or absent is made. There is no such thing as an error-free system, and the appropriate minimum output signal to noise ratio is determined by calculating mathematically the consequent probability of an erroneous decision. For many purposes a maximum error rate of 1 in 10^9 decisions is regarded as acceptable.[1]

There is virtually no thermally produced noise in the fibre. This is because the quantum of energy, hf (the minimum amount by which the energy in an electro-magnetic wave can be changed), is very large compared to the measure of available heat energy per molecule, kT, at optical frequencies. The point will be reinforced if we put appropriate numbers into the formula given in Equation (2.18) in Chapter 2. There is, however, in the received light, a form of noise consequent on the quantized nature of light known as *quantum noise*.[2]

As the light travels along the fibre it is absorbed (or reflected or scattered) by the attenuating processes in quantum sized chunks – using the photon model one would say that photons are removed from the signal. If the distance is large enough there is a finite possibility that a pulse, in which there should be light, arrives with no photons left in it, by chance they have all been removed. This outcome is equivalent to a pulse of noise just cancelling out the signal. By applying statistical techniques it is possible to show that a probability of 1 in 10^9 of this occuring arises when the average number of photons arriving per symbol interval in which there should be a pulse is about 20. Let us see what this implies in terms of ultimate receiver sensitivity.

Example calculation

What is the quantum limit for received power in a system operating at 1.3 μm wavelength and a signalling rate of 680 Mbaud (one of the standard rates)?

The energy in one photon is

$$hf = \frac{hc}{\lambda} = \frac{(6.63 \times 10^{-34}) \times (3 \times 10^8)}{1.3 \times 10^{-6}} \, \text{J} = 1.53 \times 10^{-19} \, \text{J}$$

so the energy in 20 photons $\approx 3 \times 10^{-18}$ J.

There are 680×10^6 symbol intervals per second of which, one can assume, on average half will contain pulses (representing 1s). So, the received power for an average of 20 photons per pulse is

$$(340 \times 10^6) \times (3 \times 10^{-18}) \, \text{W} \approx 10^{-9} \, \text{W}$$

In dBm this is

$$10 \log_{10} 10^{-6} = -60 \, \text{dBm}$$

The usual value of receiver sensitivity is at least 20 dB higher than this which indicates that noise generated in the receiver is the dominant effect. Current technology gets nowhere near the quantum limit.

[In the receiver, in the process of converting the light signal into an electrical signal, ideally each photon received results in an electron crossing the potential

---

[1] There exist error correcting codes which can be applied to the signal to reduce the error rate further, but that is another story.

[2] Do not confuse quantum noise with *quantization noise* which occurs when analogue signals are digitally encoded for transmission.

barrier in the photodiode (assuming that there is no avalanche gain). The *shot noise* associated with this is the quantum noise in another guise.]

Repeaters

It should be clear from what has gone before that single mode fibre links at 1.3 μm wavelength have a practical maximum span of 30 to 40 km, and hence for transmission over distances greater than this it is necessary to receive the optical signal, convert it to an electrical signal, clean up the pulses, amplify the signal and then retransmit it. This is not unreasonable over land – although it does introduce problems as we shall see next – but it is a great nuisance for transmission under oceans.

The problems caused by electrical repeating are two: first, since as we have seen noise is introduced by the receiver, each repeat introduces errors; second, when the dispersed pulses are sharpened up there is an ambiguity in the exact positions of the symbol interval boundaries – over a number of repeats this introduces what is known as *jitter* so that the final receiver is not sure where the centre of a symbol interval, the most appropriate place to make a symbol decision, is.

5.5 LONG DISTANCE SINGLE MODE PROPAGATION AT 1.55 μm

This is the wavelength at which attenuation in single mode silica fibre is least and values of 0.2 dB/km are attained. Against that, the dispersion, in standard step index fibre, is typically 17 ps/nm per km. It is possible by reducing the core diameter and adjusting the doping profile of the fibre to move the wavelength of zero dispersion to 1.55 μm, but in doing so one finds the attenuation is increased, and the fibre is more expensive.

If you construct a power budget similar to Table 5.1 for 1.55 μm signals in standard single mode silica fibre, assuming similar values for the parameters other than the fibre loss, you will see that spans in excess of 60 km are possible from an attenuation viewpoint. The dispersion, however, can only be kept within limits at high bit rates by using single mode lasers, and these are expensive and not very robust. The prospect of halving the number of repeaters compared to a 1.3 μm system may not be worth the expense, but the use of this wavelength comes into its own for trans-oceanic links because of two other phenomena which will now be described.

Soliton propagation

A soliton is a pulse that does not spread despite the dispersion effects in the fibre: this is achieved by using a non-linearity inherent in the fibre continuously to counteract the dispersion.

The refractive index of silica, besides being frequency dependent, is also slightly intensity dependent: this, as we shall see, indicates a non-linearity resulting in frequency changes in different parts of a pulse which can affect the group velocity of different parts of the pulse. The refractive index at a given frequency can be written

$$n = n_0 + n_2 I$$

where n_0 is the refractive index at near-zero signal intensity, I is the intensity (watts per square metre) and n_2 is a parameter which must have dimensions m^2/W.

The value of n_2 for silica at 1.55 µm wavelength is about $3 \times 10^{-20}\ m^2/W$. At a power level of about 1 mW in a core of diameter 9 µm the intensity is of the order of

$$\frac{10^{-3}}{\pi \times (4.5 \times 10^{-6})^2} \approx 1.6 \times 10^7\ W/m^2$$

so the change in refractive index when the power level drops to near zero is

$$(3 \times 10^{-20}) \times (1.6 \times 10^7) \approx 5 \times 10^{-13}$$

This, as you can see, is a tiny proportion of the nominal refractive index of 1.46, but none the less the effect can be used as follows.

Call the transmitted frequency – the centre frequency of the radiation in the transmitted pulses – f_t.

The phase change coefficient in the fibre

$$\beta_f = 2\pi/\lambda_f$$

where λ_f is the guide wavelength in the fibre, and

$$\lambda_f = v_p/f_t$$

where v_p is the phase velocity in the fibre. Also

$$v_p = c/n$$

so

$$\beta_f = 2\pi f_t n/c = 2\pi f_t (n_0 + n_2 I)/c$$

Two points a distance L apart in the fibre will each have oscillating electric and magnetic fields with a phase difference given by

$$\beta_f L = 2\pi f_t (n_0 + n_2 I) L/c$$

Now suppose this phase difference were changing; that would imply that the frequencies of the oscillating fields at the two points were not the same. The rate of change of the phase difference would indicate the frequency difference between the two points.

Consider a length of fibre over which the intensity of the radiation is increasing with time. It can be seen that $\beta_f L$ will also be increasing since, differentiating

$$\frac{d(\beta_f L)}{dt} = \frac{2\pi f_t n_2 L}{c} \frac{dI}{dt}$$

Now if the phase lag of the end of this section (in the direction of propagation) is increasing with time relative to the beginning, this implies that the frequency at the end of the section will be lower than that at the beginning. The change of frequency is given by

$$\delta f = -\frac{d(\beta_f L)}{dt} = -\frac{2\pi f_t n_2 L}{c} \frac{dI}{dt}$$

As a pulse, shaped perhaps like that shown in Fig. 5.3, travels along a fibre, the section of fibre which the front end of the pulse is entering experiences an increase of intensity with time, while the section which the rear end is leaving experiences a reduction of intensity with time, hence effectively the radiation in the front end of the pulse has a reduced frequency while that in the rear end has an increased frequency. Over the wavelength range centred on 1.55 μm the group velocity increases with frequency, so the non-linear effect tends to push the pulse together and counteract the effect of (linear) dispersion. By carefully shaping the pulses and selecting the appropriate power level, and by manipulating the dispersion of different sections of the fibre (by means of its structure) it has proved possible to propagate soliton pulses over very large distances.[3]

Signalling rates of 5 Gb/s (5×10^9) are routinely used and before the year 2000 will almost certainly have been exceeded. The symbol interval corresponding to 5 Gb/s is 200 ps. Because soliton propagation can easily fail if the pulses are allowed to overlap, each pulse normally occupies only one-fifth of the symbol interval – i.e. 40 ps at 5 Gb/s. It is also necessary to prevent reflection from the receiver, and for this purpose special isolators are used – these will be dealt with briefly later. There remains, however, one outstanding problem; how to maintain the power level without needing to insert electronic repeaters (which would defeat the purpose): this can be achieved by using erbium-doped fibre amplifiers.

EDFAs

The erbium-doped fibre amplifier is a close relative of the laser, depending as it does on stimulated emission from erbium atoms in the fibre to enhance the signal amplitude.

The electrons in any isolated atom have distinct energy levels: these are broadened into energy bands when the material forms a solid. When a small

---

[3] It should be noted that the spread of frequencies in the pulsed output of a typical laser source varies through the pulse being higher at the rising edge than at the falling edge – the opposite of the effect produced by the non-linearity. This must also be taken account of in designing the system.

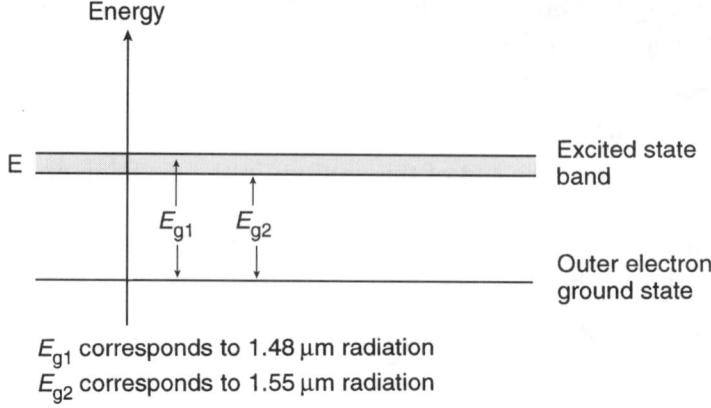

E_{g1} corresponds to 1.48 μm radiation
E_{g2} corresponds to 1.55 μm radiation

Fig. 5.4   Part of the energy level diagram for erbium atoms in silica

proportion of erbium is introduced into a silica fibre the individual erbium atoms are not very close to each other so the energy levels of the electrons in the unexcited atoms are not significantly broadened. However, orbitals which an outer electron of the atom will occupy if the atom is excited are broadened to some extent. Figure 5.4 illustrates part of an energy level diagram for erbium atoms in this condition; the important feature is the narrow band of energy levels marked E. The lowest level in this band represents a *metastable state*, which means that an electron, when it finds itself in this state, stays there for about 14 ms (a very long time in electron-transition terms) before it falls back to its ground state, and when it does so, it emits a photon of radiation of a frequency consistent with the change in energy according to the formula $hf = \delta E$. The wavelength associated with the transition from the metastable state to the ground state in erbium is 1.55 μm which is why the element is so useful in this application.

The energy difference between the electron's ground state and the top of the energy band E is that of photons of a wavelength of 1.48 μm, so light of this wavelength can be used to 'pump' the amplifier.

Figure 5.5 is a representation of the arrangement. The doped fibre in the loop is about 20 m long. The main signal, consisting of pulsed 1.55 μm radiation in the

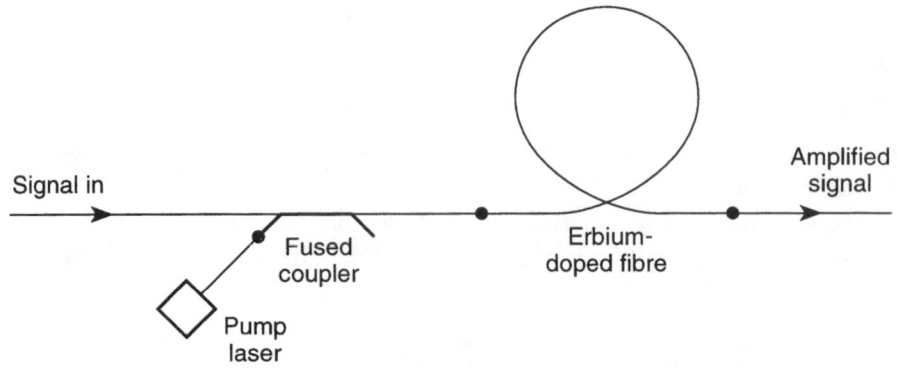

Fig. 5.5   An erbium-doped fibre amplifier

transmission fibre, has added to it continuous-wave radiation of wavelength 1.48 µm from a 'pump laser' by means of a *fused-fibre coupler*, a device which allows a second signal to be introduced into a fibre without the main signal leaking out. The 1.48 µm radiation excites an electron in many of the erbium atoms to the top of the excited band whereupon those electrons rapidly lose energy to the fibre in the form of heat and 'trickle down' to the metastable level at the bottom of the band. When a photon of the signal radiation passes an erbium atom in its ground state the photon is absorbed and the erbium atom is put in its excited state, but if the erbium atom is in the excited (metastable) state it is induced to make its transition to the ground state and emit a photon which is in phase with the stimulating photon. So long as there are more atoms in the metastable state than in the ground state the signal level will be enhanced. Of course, some of the electrons in the metastable state will make random, as opposed to stimulated, transitions and this gives rise to optical noise, but because of the long duration of the metastable state when undisturbed, the level of this is very low. The random emission noise, however, is amplified in its subsequent journey through the doped fibre, just as the signal is, so the consequent noise is known as amplified spontaneous emission (ASE) noise.

There are, of course, any number of variable design details, some of which are as follows. It is possible to use a pump signal at a shorter wavelength – 980 nm – because there exists another excitation energy level at the appropriate separation from the ground state, which is not metastable and from which an excited electron descends rapidly to the metastable level. However, pump energy at this wavelength will be attenuated more rapidly by the fibre. The pump signal can be fed in from either end of the doped section: pumping in the direction of the signal produces less noise, but pumping the other way reduces the likelihood of instability due to reflections. It may be found necessary to include an isolator at one or both ends to prevent the device from oscillating.

A gain of 20 dB with a bandwidth of 30 to 35 nm is typically achieved.

The production of amplifiers of this sort to work at other signal wavelengths depends on finding materials with the appropriate metastable energy levels: some success is reported at 1.3 µm signal wavelength using praseodymium in silicon fluoride fibre.

Optical isolators

If a linearly polarized electromagnetic wave is passed through a transparent material of high refractive index along the direction of a magnetic field, the direction of polarization is rotated. The sense of rotation depends on the direction of the magnetic field, not on the direction of propagation of the radiation. This effect, known as Faraday rotation, is used in optical isolators in the following way. In the HE_{11} mode in single mode propagation the electric field is in one direction – similar to a linearly polarized plane wave. At the input end of the isolator the direction of polarization is established by passing the signal through a layer of a polaroid type of material. The length of the isolator and the strength of the magnetic field are adjusted to produce 45° of rotation of the polarization. Any reflected signal is rotated through a further 45° in the same sense in passing back through the isolator and is therefore absorbed by the polaroid.

5.6   MULTIMODE FIBRE PROPAGATION

The attenuation in multimode fibre should be much the same as in single mode, assuming the same quality of materials and manufacture, so from that point of view it would be best to use radiation of wavelength 1.55 μm for transmission over any significant distance. A number of current multimode systems use 1.3 μm radiation, but that is just because signal sources at that wavelength were more readily available when they were designed.

Intermodal dispersion in step-index fibre

The effect represented by intermodal dispersion is that copies of a pulse launched on the fibre will propagate in the different modes with different group velocities and so arrive at the receiver at different times. The dispersion over a distance L can be represented simply by the difference in transmission time taken by a pulse in the fastest mode and a pulse in the slowest mode. Thinking about the progress along the fibre of the generating wave in different modes it should be clear that the fastest mode will be the fundamental mode. The slowest mode will be the one for which the generating wave strikes the core boundary the most obliquely; in fact, at the critical angle, which is the most oblique angle at which the generating wave can strike the boundary and be reflected.

Assuming a typical core diameter of 50 μm, the generating wave associated with the fundamental mode will propagate almost along the axis of the core (think of the propagation of the fundamental mode in metal waveguide far above its cut-off frequency). The condition, related to Fig. 5.1, is on the far right of the graph: the evanescent fields do not penetrate very far into the cladding and the average refractive index encountered by the mode is effectively n_1. So, the group velocity of this mode is effectively c/n_1 and the time taken by a pulse in this mode to travel a distance L in the fibre is

$$T_1 = Ln_1/c$$

Figure 5.6 shows the propagation direction of a generating wave reflecting from the core boundary at the critical angle. The group velocity of this mode can be taken as the rate of progress of generating wave energy along the fibre core, i.e.

$$v_{gm} = \frac{c}{n_1}\cos\theta = \frac{c}{n_1}\sin\phi$$

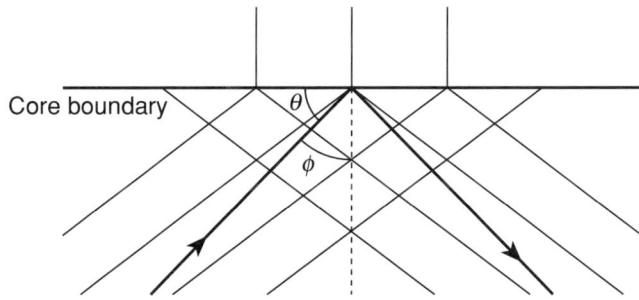

Fig. 5.6   The generating wave reflecting at the critical angle from the core boundary

Since ϕ is the critical angle, $\sin \phi = n_2/n_1$ so

$$v_{gm} = \frac{cn_2}{n_1^2}$$

The time taken by a pulse in this mode to travel a distance L in the fibre is

$$T_2 = \frac{Ln_1^2}{cn_2}$$

The intermodal dispersion over a distance L is therefore

$$T_2 - T_1 = \frac{Ln_1^2}{cn_2} - \frac{Ln_1}{c} = \frac{Ln_1}{cn_2}(n_1 - n_2)$$

Again it is desirable to make the difference between the core and cladding refractive indices as small as other considerations will allow; taking $n_1 = 1.460$ and $n_2 = 1.458$, the above formula gives the intermodal dispersion over a length of 1 km as 6.68 ns.

Graded-index fibre

Intermodal dispersion can be reduced in multimode propagation by using a fibre with a core refractive index that varies from a maximum in the centre to equal that of the cladding at the interface – *see* Fig. 5.7.

There is no sudden boundary; the generating wave associated with a given mode is contained by bending back rather than sudden reflection. The velocity of propagation is higher near the cladding than at the centre, so generating waves which suffer many reflections (associated with higher-order modes) spend more time in the high-velocity region. *See* Fig. 5.8, which illustrates the 'rays' of the generating waves associated with two different modes. This has the effect of reducing the difference in group velocity between the different modes, and hence reduces the intermodal dispersion relative to a step-index fibre. Detailed analysis is complex but the outcome is that the optimum profile for the refractive index in the core is a parabola. Intermodal dispersion of around one-tenth of the value for the equivalent step-index fibre is achieved.

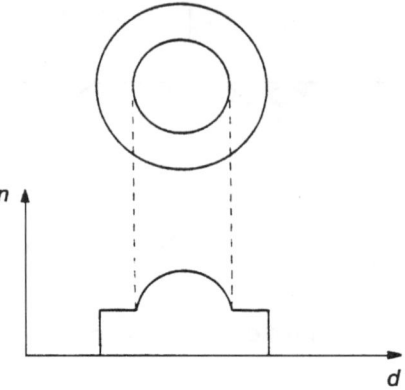

Fig. 5.7 Refractive index profile of graded-index optical fibre

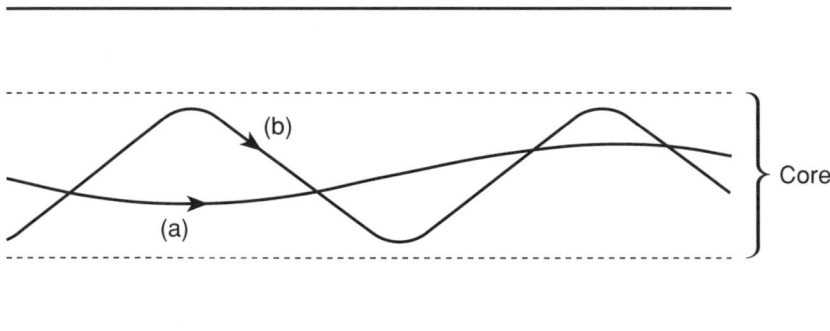

Fig. 5.8 Propagation of two different modes in graded-index optical fibre, (a) ray for lower-order mode, (b) ray for higher-order mode

5.7 MECHANICAL CONSIDERATIONS

Bends

If the fibre has to be bent, this will obviously affect the angle at which the generating wave meets the core boundary, and may result in light leaking from the fibre. There is also the danger that the material of the fibre will develop cracks due to the bending stress. The manufacturer will specify the minimum radius of bending that can safely be used, bearing in mind both of these constraints. When fibres are assembled in cables, care has to be taken to avoid stressing and/or kinking the fibre. Attenuation caused by bending may need to be included in the power budget.

Fibre termination

The problem of terminating a fibre at each end with the minimum reflection is largely a mechanical one. As in the case of waveguide it would in theory be possible to define a characteristic wave impedance for the fibre, and use that to design the termination, but this is not a practical approach. In practice the design of the fibre termination at the receiving end is incorporated in the design of the photodiode, and its quality is judged by the size of its reflection coefficient.

5.8 LINK CALCULATIONS

Imagine that you require a step-index single mode fibre for 0.85 μm radiation.

5.1 Choose a suitable core diameter, assuming the same core and cladding refractive indices as described for 1.3 μm fibre.

5.2 Taking the attenuation at this wavelength as 1.5 dB/km and assuming suitable values for splice losses and connector losses, estimate the maximum span possible between a transmitter of power −3 dBm and a receiver of sensitivity −44 dBm, and draw up a link power budget.

5.3 Assuming that you wish to use a signalling rate of 36 Mbaud and that the dispersion of the fibre is 60 ps/nm per km, estimate the maximum source linewidth to achieve a span as indicated in the previous calculation.

5.4 Calculate what span would be achievable using the same wavelength and signalling rate in multimode fibre and show that a low power LED of, say, −13 dBm would be more than adequate as a signal source.

6 Free space

The medium that we are concerned with here, although called 'free space', includes the atmosphere. The discussion will include the general principles of antennas, and two examples of links using microwave radiation (with wavelengths in the centimetre range), to illustrate the considerations involved.

All emissions of electromagnetic radiation are regulated by international agreement; the adherence to regulations is enforced by law within individual countries. Allocations are made according to the purpose of the emission, and include maximum power to be emitted as well as frequencies and the accuracy with which they are to be maintained.

In principle, a given purpose will often dictate an appropriate frequency – for example at very low frequencies radio waves diffract round the earth and so can be used for long-distance telegraphy. In practice this is not by any means the only criterion; there is a great deal of 'politics' in the allocation of frequency bands.

Because of the demands on the spectrum, the same frequencies have to be used for different transmissions; there are three different ways in which this can be successfully done. In the first, the distance separating service areas using the same frequency is chosen to be sufficiently large for signals from one area to cause an acceptably low level of cross-talk in the other. The second method of frequency reuse is to use highly directional antennas so as to restrict the radiation to relatively narrow beams. The third is to use orthogonal polarizations (usually horizontal and vertical polarization).

6.1 ANTENNAS

A *transmitting antenna* is a device for launching an electromagnetic wave; a *receiving antenna* extracts an amount of power from an incident electromagnetic wave that is proportional to the wave's power density.

Although transmitting antennas and receiving antennas seem to have qualitatively different functions they have many common properties, and, in fact, the same structure can often be used for either purpose. This general principle is known as *reciprocity*.[1]

---

[1] Some modern antennas contain arrays of active devices (amplifiers and phase shifters for instance). Such an antenna cannot be changed from receiving to transmitting and vice versa without modification.

Generally, an antenna's size is related to the wavelength of the signal that it is designed to radiate or receive, so any given antenna can only be used for a restricted range of frequencies. Antennas for low frequencies are very large and usually need to be mounted near the earth's surface – in wavelength terms. The earth (or sea) acts as a reflecting plane and has to be included in the antenna design; antennas used for 'long wave' and 'medium wave' broadcasting are a case in point. The earth's surface does not affect the performance of antennas for higher frequencies than this although it may reflect the waves being transmitted or received. The design of antennas derives from electric circuit theory at low frequencies and optics at high frequencies with a gradual transition between the two approaches as the design frequency increases.

Directional properties

A transmitting antenna is designed to have specified directional properties, that is, to radiate power in desired directions and not in undesired directions. Thus we may have a broadcast antenna, designed to radiate in all directions along the earth's surface, but not vertically upwards, or we may have a highly directional antenna which concentrates nearly all the radiated power into a narrow cone. If the same structure is used as a receiving antenna, it is found that its sensitivity has exactly the same distribution, i.e. it absorbs power coming from the desired directions, but not from the undesired directions. This is an example of reciprocity.

The directional properties of an antenna can be conveniently displayed on a *polar diagram*, that is, a polar plot of transmitted field strength (or received amplitude) against direction at a fixed distance. A true polar diagram for an antenna is three-dimensional, but usually appropriate two-dimensional cross-sections are plotted.[2]

The *directivity*, *D*, of a transmitting antenna is defined as the ratio of the power density, at a given distance, in the direction of maximum radiation, to the power density that would occur at the same distance if the same total power were transmitted by an *isotropic antenna* (one radiating equally in all directions). A closely related quantity is the *antenna gain*, *G*. This is defined as the ratio of the maximum power density from the given antenna compared to that from a loss-free isotropic radiator fed with the same transmitter power. The gain is less than the directivity by an *efficiency factor* determined by ohmic losses in the antenna. The gain is the quantity that can be measured and in much of the literature ohmic losses are implicitly ignored and 'gain' is referred to when directivity is what is really meant. Note that the antenna gain does not increase the total radiated power; in that sense it is a misleading term.

Directivity and gain are also specified for a receiving antenna. In this case they are defined in terms of received power from the best direction compared to received

---

[2] There is a problem of confusing terminology here; the polar diagram has nothing to do with the direction of polarization of the wave, which indicates the direction of the electric field in the wave.

power at the terminals of an isotropic receiving antenna. However, the gain has most probably been measured, or calculated, for the same antenna used as a transmitter and assumed to be the same in the receiving mode (reciprocity again).

Radiation resistance

Any radiating structure that is fed electrically will present to the transmitter an impedance, the real part of which accounts for power radiated (assuming that ohmic losses are negligible). If the antenna is designed so that this impedance is purely real (as is usually the case) it is called the *radiation resistance*. The same antenna, receiving, acts as a signal source with an internal impedance which is this same radiation resistance (reciprocity yet again). Knowledge of the radiation resistance is important because of the need to match the antenna to its feed and in the calculation of noise in a received signal.

Effective aperture

The effective aperture of a receiving antenna is the area of the wavefront from which the antenna can be deemed to have extracted power. Thus, if an antenna delivers to a receiver 2 milliwatts of power and the wave has a power density of 1 milliwatt per square metre, the antenna effective aperture must be 2 square metres (notice that any ohmic losses in the antenna are implicitly included in this definition). Aperture also has significance when the antenna is used for transmission, as we shall see, and can be related to the antenna gain.

Primary radiators

A half-wave dipole is probably the simplest efficient radiating structure. (There does not exist a simple structure which radiates isotropically, although the concept is useful for defining things like directivity.) It is a rod, approximately $\lambda/2$ long, divided and fed at the centre and radiating a wave polarized in the direction of its length. Insight into its function can be gained, as suggested in Chapter 2, by thinking of it as a bent-out twin-wire transmission line.

The detailed theory of half-wave dipoles gives the following results:

- The bandwidth over which the input impedance is effectively resistive depends directly on the dipole thickness (that is, the thickness of the rod from which it is made).
- The *resonant length* (for resistive input impedance) is exactly $\lambda/2$ only if it is infinitely thin: the resonant length gets progressively less as the dipole is made thicker.
- The directivity of a half-wave dipole is 1.64 (≈ 2 dB). This is a theoretical result which allows the gains of other antennas to be measured by comparing their directive properties with that of a half-wave dipole.

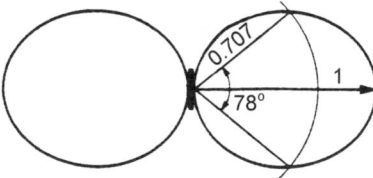

Fig. 6.1   Polar diagram of a half-wave dipole in the plane of the dipole

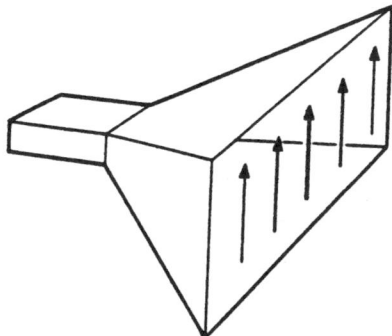

Fig. 6.2   A rectangular waveguide horn (arrows represent the **E** field)

- The polar diagram, in a plane containing the antenna, is shown in Fig. 6.1. In a plane perpendicular to the antenna it is a circle. (The three-dimensional polar diagram is doughnut shaped.)

Another important primary radiator is the *waveguide horn*. The open end of a waveguide, carrying the fundamental mode, does not act as an electrical open circuit; in fact it produces a standing wave ratio of around 3:1, which indicates that a large proportion of the power is radiated from the end. The waveguide can be 'matched to free space', so that all the power is radiated, by flaring it out at the end into a horn as shown in Fig. 6.2. The direction of polarization of the radiation is the same as in the waveguide that feeds the horn. The radiation from a horn is directional: the general shape of the polar diagram is shown in Fig. 6.3. This form of polar diagram results from any radiating aperture, and will be considered in a little more detail later. It has a *main lobe* and *side lobes*. The bulk of the power is in the main lobe. The angle of spread of the main lobe decreases as the dimension across the exit of the horn increases.[3]

Antenna arrays

A row of dipoles, fed in phase, is much more directive than the individual dipoles. Their individual radiations interact by the process of *interference* to give a desired composite pattern. An example is shown in Fig. 6.4: six dipoles are shown with

---

[3] Some of the microwave antennas which can be seen on telecom towers are modified waveguide horns.

Fig. 6.3   Form of the polar diagram of a waveguide horn

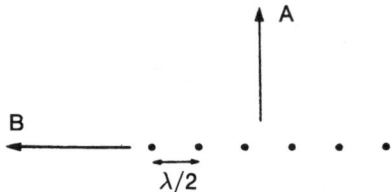

Fig. 6.4   Cross-section of a broadside array of dipoles

their lengths into the page. If they are fed in phase, their radiation reinforces in the direction A, but cancels in direction B; hence this is called a *broadside array*. The array has a polar diagram, in the plane of the page, as shown in Fig. 6.5. In a plane along direction A and perpendicular to the paper, the polar diagram is that of an individual dipole.

A metal reflector, placed behind the array at a distance of $\lambda/4$, turns one lobe round, to give a polar diagram as shown in Fig. 6.6. Radiation which travels to the reflector perpendicular to the plane of the array and is reflected back has a path-length of $\lambda/2$ plus phase inversion on reflection, and so is in phase with the direct forward radiation. In the direction of maximum intensity the field strength is doubled, so that the power density is increased four-fold, indicating that the gain is increased by 6 dB.

Linear arrays can be stacked on top of each other, to give more directivity in the second plane, as shown in Fig. 6.7.

The array shown in Fig. 6.8 is known as a *Yagi* array. Its polar diagram has a maximum in one direction along the centre line joining the dipoles: for this reason

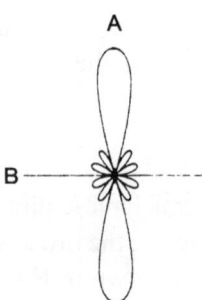

Fig. 6.5   Polar diagram of the array of Fig. 6.4, in the plane of the cross-section

Fig. 6.6   Polar diagram of the array of Fig. 6.4, with a metal reflector added

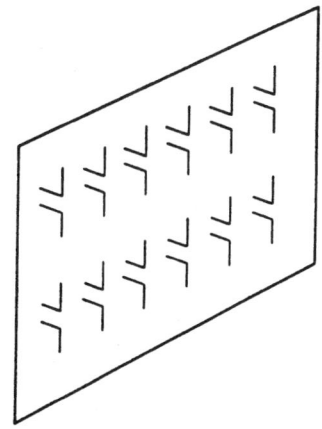

Fig. 6.7   Two-dimensional array of dipoles

it is called an *end-fire* array. It is instructive to consider qualitative explanations of its behaviour as a transmitter and as a receiver, bearing in mind that it has the usual property of reciprocity.

The directional properties of the Yagi array as a transmitter arise as follows. The dipole connected to the transmitter is known as the *driven element*; the other dipoles are *parasitic* and pick up power from the driven element. The parasitic elements are not cut to the resonant length, and the result of this is that they reradiate received energy with a change of phase. The lengths and separations of the parasitic elements are chosen so that they reradiate with phases such that their radiations reinforce along the line of the array. In all other directions there is a varying degree of destructive interference.

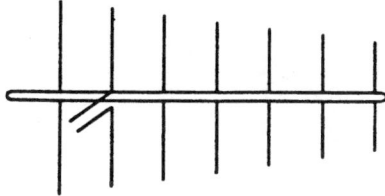

Fig. 6.8   A Yagi array

When the Yagi is used as a receiving antenna, radiation arriving along the line of the array has different phases at the various dipoles. The changes of phases on reradiation from the parasitic elements are such that the reradiated waves all arrive in phase at the connected dipole. For radiation arriving from any other direction the arrival phases are not the same as above, so the phases of the reradiated waves at the connected dipole are not optimal.

Arrays can be made from any type of primary radiator, in principle, and combine the directivity of the individual radiators with that caused by the array.

A radiating aperture

It is convenient to consider as an ideal radiating structure one which would produce a wave having a uniform phase and power density, but over a limited area, as if a plane wave were coming through a hole. Such a wave spreads out, inevitably, by the process of diffraction, but the spread is less than it would be with any other distribution of phase and power density over the same area.

The diffraction pattern produced by such an aperture at a distance large compared to its width results in a polar diagram which is evaluated in Appendix 5.

The result is by now a familiar shape; it is shown in Fig. 6.9. The angle of the main lobe proves to be simply related to the dimensions of this ideal aperture. The angle between half-power directions – marked α in Fig. 6.9, and often called the *beamwidth* – is approximately equal to the inverse of the diameter, in the same plane, of the aperture in wavelengths; i.e.

$$\alpha_{\text{radians}} \approx \frac{\lambda}{a}$$

The directivity of this radiating aperture is related by the theory to its area by

$$D = \frac{4\pi A}{\lambda^2}$$

That this formula for directivity is consistent with that for beamwidth can be illustrated as follows. Consider an ideal square-aperture antenna of side z metres. Ignore side lobes, and imagine that it radiates solely into a square cone, as shown in Fig. 6.10. If the side of the aperture is z, then

$$\alpha_{\text{radians}} = \lambda/z = \lambda/\sqrt{A}$$

where A is the aperture area. Also

$$x = \alpha R$$

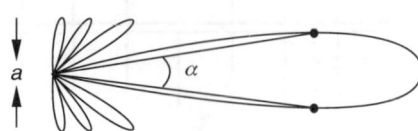

Fig. 6.9   Polar diagram of an aperture

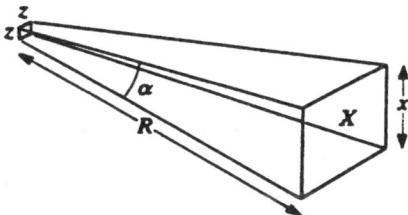

Fig. 6.10   Idealized radiation distribution from a square aperture

so the area

$$X = x^2 = \alpha^2 R^2 = \frac{\lambda^2 R^2}{A}$$

The area of the total sphere at radius R is $4\pi R^2$. The power which, if radiated isotropically, would pass uniformly through the whole sphere, is concentrated into the area X, so the directivity will be the ratio of the two areas, i.e.

$$D = \frac{4\pi R^2}{\lambda^2 R^2 / A} = \frac{4\pi A}{\lambda^2}$$

Practical aperture antennas

The mouth of a waveguide horn is a radiating aperture; however, it is not ideal because neither the phase nor the amplitude of the wave are uniform over the aperture. The effect of these imperfections is to make the half-power angle of the main lobe greater than it would be for an ideal aperture of the same area. We can specify for the horn an *effective aperture*, less than its physical aperture, which could be estimated by measuring the beamwidth, but which is more reliably obtained by measuring the gain.

We can write

$$G = \frac{4\pi A'}{\lambda^2}$$

where the value of A', the effective aperture, includes the efficiency factor of the antenna.

It is not surprising to learn that the effective aperture so defined for any transmitting antenna is identical to its effective aperture, as previously defined, when it is used as a receiver.

The other important aperture antenna which we shall discuss is the *parabolic dish* (paraboloid), illustrated in Fig. 6.11.

A paraboloidal reflector has the geometrical property that all rays of radiation coming from the focus, F, and striking the paraboloid are reflected parallel to the principal axis. Furthermore, all path-lengths, such as FX, are the same. Hence, a primary radiator (dipole or waveguide horn) placed at F will 'illuminate' the paraboloid and produce an approximation to a plane wave at the plane of its rim. The

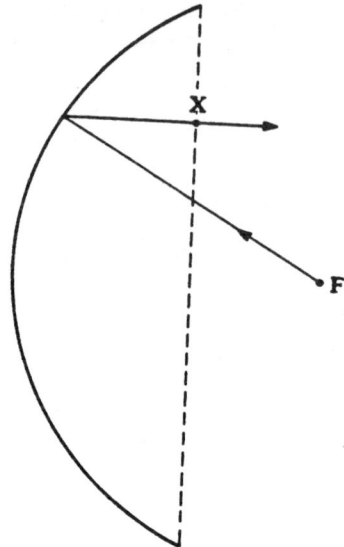

Fig. 6.11   Cross-section of a parabolic dish

illumination is not generally uniform – there is usually greater power density at the centre than at the rim – and there may be phase differences across the aperture.

A well-designed parabolic dish and feed (primary radiator) usually has an effective aperture of about 2/3 its physical aperture. The beamwidth is given approximately by

$$\alpha = \lambda/d_e$$

where d_e can be thought of as the diameter of the effective aperture, and is about 80 per cent of the actual diameter.

The polarization of radiation from a paraboloidal antenna is determined by the primary radiator.

Significant antenna properties summarized

Radiation resistance: the input resistance of the antenna in transmission and its source resistance in reception.

Directivity: the same for transmission and reception, with a polar diagram that generally has a main lobe and (undesired) side lobes.

Effective aperture: indicates the area of wavefront intercepted as a receiving antenna.

Gain: determines the degree of concentration of power flux into a beam as a transmitting antenna.

Beamwidth (angle between half-power directions in the main lobe in transmission): can be estimated using the formula $\alpha \approx \lambda/d_e$.

Relationship between gain and effective aperture:

$$G = \frac{4\pi A'}{\lambda^2}$$

6.2   MICROWAVE LINKS

Microwave links are allocated frequencies in the GHz range so that the wavelength of the radiation is a few centimetres. Two highly directional antennas point at each other and are mounted at a sufficient height to avoid obstruction by the intervening terrain. This needs a little further explanation. First of all the curvature of the earth has to be taken into account. Second, any obstruction cannot be allowed to come near enough to the line of sight to reflect any of the beam in such a way as to cause interference with the direct beam. To satisfy this second criterion it is normally taken that the path-length via any reflection must exceed the direct path-length by at least $\lambda/2$. Figure 6.12 shows the locus of all such points, and it can be seen that the required clearance is greatest around the midpoint between the antennas. At the midpoint, using Pythagoras' theorem, the requirement leads to

$$x^2 > (d/2 + \lambda/4)^2 - (d/2)^2$$

which, since $d \gg \lambda$, yields

$$x > \text{approx} \tfrac{1}{2}\sqrt{\lambda d}$$

To put this in perspective, if d is 50 km and λ is 5 cm, then x must be greater than $\tfrac{1}{2}\sqrt{0.05 \times (5 \times 10^4)} = 25\,\text{m}$.

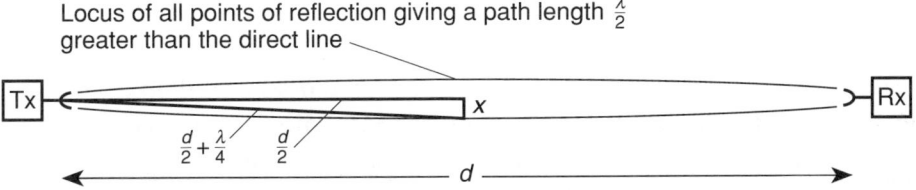

Fig. 6.12   A microwave link

Received power

Consider a transmitter and a receiver, with antenna gains G_T and G_R respectively, separated by a distance d, as shown in Fig. 6.12. If the transmitter power is P_T watts, then, assuming no absorption along the path, the power flux density at the receiver antenna is

$$\frac{P_T G_T}{4\pi d^2}$$

In transmitter specifications, the product $P_T G_T$ is often lumped together and quoted as the *equivalent isotropic radiated power* (EIRP). This is represented by the symbol P_{ei}.

Losses can be introduced into the equation as a factor L (less than unity): this will account for atmospheric absorption and may also include the effect of *pointing errors* – a reduction in the received power caused by the two antennas not accurately pointing at each other. So now we can write for the power flux density at the receiver

$$\frac{P_{ei} \cdot L}{4\pi d^2}$$

The greater the loss the smaller the value of the loss factor L.

The power into the receiver will be this expression multiplied by the effective aperture of the receiver.

Transforming the formula

$$G_R = \frac{4\pi A'}{\lambda^2}$$

The effective aperture of the receiving antenna is

$$A' = \frac{G_R \lambda^2}{4\pi}$$

The power of the signal entering the receiver is denoted C. C is used because at this stage the signal is a modulated carrier. We can write

$$C = \frac{P_{ei} \cdot L}{4\pi d^2} \cdot \frac{G_R \lambda^2}{4\pi}$$

and then rearrange the terms on the right hand side to give

$$C = P_{ei} \cdot L \cdot \left[\frac{\lambda}{4\pi d}\right]^2 \cdot G_R$$

The term $[\lambda/4\pi d]^2$ is called the *free space path loss*. It does not result from absorption, but from the dilution of the power flux density with distance: as we can see, it is proportional to the inverse square of the transmission distance in wavelengths.

Writing L_P for the free space path loss

$$C = P_{ei} \cdot L \cdot L_P \cdot G_R \tag{6.1}$$

Noise spectral power density

The noise that corrupts the received signal comes from three sources. Some of it is generated in the transmitter and transmitted with the signal. Some of it enters the receiver antenna from the surroundings: if the antenna has a high gain, then most of this comes from where the antenna is pointing (some comes in via side-lobes) and it is often called *sky noise*. The rest of the noise is generated in the receiver itself.

Most of the standard theory of noise is based on the assumption that the noise is unavoidable and has statistical properties described as additive, white and gaussian. Cross-talk and interference may not satisfy these requirements, but, nonetheless, they are often lumped in with natural sky noise (which does) in an analysis.

In Chapter 2, a formula, Equation (2.18), was given for thermally generated noise delivered by a medium to a receiver. The photon energy at microwave frequencies is not significant in this formula at normal ambient temperatures. The extra noise that is not thermal is also proportional to bandwidth, so we can still use the formula kTB for delivered noise if we use a value for T which is not the actual temperature, but a fictional higher temperature. Under these circumstances it is convenient to include also noise generated in the receiver to get a measure of the output noise. This receiver noise is calculated 'as if' it occurred at the receiver input (i.e. discounting the extra gain to the point where it actually occurs) known technically as 'referring the noise to the receiver input'. The total effective noise power at the input of the receiver is then represented as kT_eB where T_e is an *equivalent noise temperature*. In order to compare behaviour for different bandwidths it is convenient to work in terms of the noise power per unit bandwidth or *noise spectral power density* (often abbreviated to simply *noise density*),[4] N_0, for which we can write

$$N_0 = kT_e \tag{6.2}$$

It is normal practice to ignore any transmitter noise and assume that the equivalent noise temperature at the input of the receiver represents only sky noise plus noise generated in the receiver, so we can write

$$T_e = T_s + T_R$$

For a high gain receiving antenna the equivalent noise temperature of sky noise is largely determined by the actual temperature of the region at which the antenna is pointing, thus for an antenna directed along the earth's surface T_s will be around 290 K.

Carrier to noise density ratio

From the results of the last two subsections, dividing Equation (6.1) by (6.2), we can now write

$$\frac{C}{N_0} = P_{ei} \cdot L \cdot L_P \cdot \frac{1}{k} \cdot \frac{G_R}{T_e} \tag{6.3}$$

For a microwave link, the carrier to noise density ratio at the receiver output just before the pulses are demodulated is an essential parameter. The probability of

---

[4] It is unfortunate that the word 'density' appears in two different contexts – in the power (flux) density of a wave, which is the average power flux through unit cross-sectional area, and in noise (spectral power) density, which is the noise power per unit bandwidth. This ambiguity should not, however, cause confusion so long as one is aware of it.

error in the demodulated pulses depends on the carrier power to noise power ratio at the demodulation point, so C/N_0 indicates how much bandwidth, and so what pulse rate, is possible.

Notice again the way the terms have been arranged. $1/k$ is a constant. G_R/T_e is a figure of merit for the receiver and its antenna and is known as the *gain/noise-temperature ratio*. (Sometimes this is abbreviated to gain/temperature ratio.)

To summarize the terms on the right hand side of Equation (6.3), we have

- P_{ei}, the equivalent isotropic radiated power (EIRP). This is the product of the power into the transmitting antenna and its gain.
- L, the attenuation loss. This represents losses in the atmosphere together with any power reduction caused by the antennas not pointing at one another accurately.
- L_p, the path loss. This is given by the formula $(\lambda/4\pi d)^2$ where d is the path length.
- $1/k$, the inverse of Boltzmann's constant.
- G_R/T_e, the receiver gain temperature ratio. This is the ratio of the gain of the receiving antenna and its equivalent noise temperature.

Example calculations

Consider a link of length 50 km between parabolic dishes, each of diameter 1.2 m. The transmitter power is 2 W and the carrier wavelength 6 cm. The receiver noise temperature is 600 K.

The effective aperture of each antenna can be estimated as 2/3 its physical aperture, i.e.

$$2/3 \times \pi \times 0.6^2 = 0.75\,\text{m}^2$$

This indicates a gain,

$$\frac{4\pi A'}{\lambda^2} = \frac{4\pi \times 0.75}{0.06^2} = 2618$$

so

$$P_{ei} = P_T G_T = 2 \times 2618 = 5.24 \times 10^3 = 37.2\,\text{dB}$$

We have to make an estimate for L. Based on experience of the worst possible signal absorption by rain, and an allowance for pointing error we take (worst case)

$$L = -6\,\text{dB}$$

$$L_p = [0.06/4\pi \times (50 \times 10^3)]^2 = 9 \times 10^{-15} = -140.4\,\text{dB}$$

$$k = 1.38 \times 10^{-23}\,\text{J/K}$$

so $1/k$ in 'dB' $[10\log(1/1.38 \times 10^{-23})] = 228.6\,\text{dB}$

$$T_e \approx 290\,\text{K} + T_R = 890\,\text{K}$$

so, $G_R/T_e = 2618/890 = 2.94 = 4.7\,\text{dBK}$.

We can now set out a power budget based on Equation (6.3)[5]

EIRP	37.2 dBW
Atmospheric absorption	-6 dB
Free-space path loss	-140.4 dB
$1/k$	228.6 dB
G/T	4.7 dBK
C/N_0	124.1 dBHz

Suppose the minimum acceptable carrier to noise ratio at the demodulator is 36 dB. This allows a bandwidth equivalent to

$$124.1 - 36\,\text{dB} = 88.1\,\text{dB} = 6.4 \times 10^8\,\text{Hz}$$

indicating that the link could operate at a bit rate up to 640 Mbits/s.

Free space is not dispersive, although any reflections might cause an effect equivalent to intermodal dispersion in fibre.

6.3 SATELLITE TRANSMISSIONS

Transmissions to and from satellites have to be at frequencies determined by ease of penetration of the ionosphere and of the atmosphere by the radiation, and also by the size of antennas that can reasonably be mounted on the satellite. These considerations dictate the use of microwaves.

Circular polarization

When sending and receiving linearly polarized signals to and from a satellite from different points on the earth's surface it is very difficult to ensure that the receiving antenna is correctly orientated with respect to the transmitted direction of polarization. This problem is made worse by the fact that Faraday rotation sometimes occurs in the ionosphere. For these reasons it is common practice to transmit waves with a form of polarization known as circular polarization, which will now be described.

Suppose that we wish to receive a linearly polarized electromagnetic wave with a dipole, but cannot predict the direction of polarization of the wave relative to the line of the dipole. If the dipole is in the direction of polarization of the wave we shall receive the maximum signal, whereas, if the dipole is at right angles to the wave's polarization we shall receive effectively zero signal. If the two are at an angle other than 90°, then the dipole will pick up the component of the wave resolved so that its electric field is parallel to the dipole. At first

---

[5] Notice the different lay-out from that of the power budget for a fibre link. This is partly because the two technologies developed separately, but mainly because noise from the fibre is negligible, so the outcome is related to a different criterion.

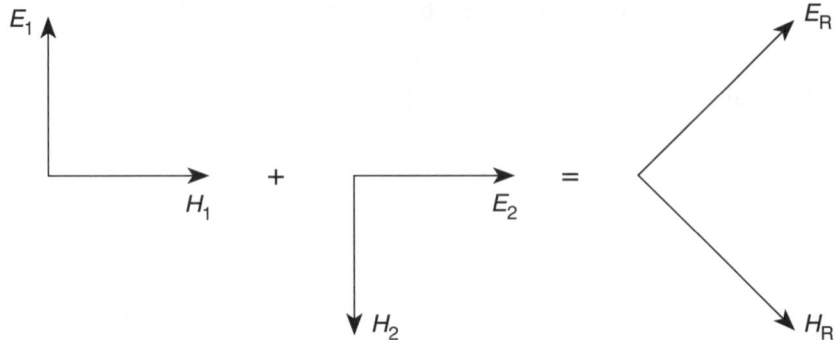

Fig. 6.13   Two linearly polarized waves (propagating into the page) combine to form another linearly polarized wave

sight it would seem that the dipole signal could be made the same for all orientations by radiating exactly similar waves with two directions of polarization at right angles, but this will not work because the fields of the two waves add together vectorially to produce another plane polarized wave, as shown in Fig. 6.13.

The problem can be overcome by radiating two waves of equal amplitude (and frequency) with directions of polarization at right angles, but also in phase-quadrature. Two such waves do not combine to produce another linearly polarized wave. A dipole aligned with one or other will pick up energy from the wave with which it is aligned; for other orientations it will pick up energy from the resolved part of each.

What is the nature of the composite wave that results? In Fig. 6.14 are drawn vectors to represent the electric fields of the two waves, in a single plane, at times t_1, t_2, t_3, t_4, over a quarter cycle. Since the waves are in phase quadrature, we can write (taking t_1 as the zero of time)

$$e_1 = E \cos \omega t \qquad \text{and} \qquad e_2 = E \sin \omega t$$

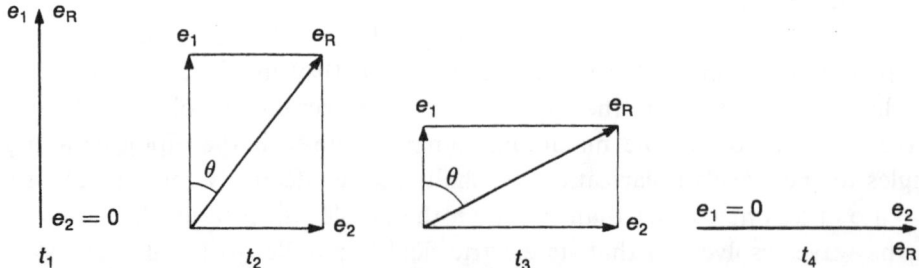

Fig. 6.14   Combination of two linearly polarized electromagnetic waves to form a circularly polarized wave. Electric fields only are shown: magnetic fields combine similarly at right angles to the electric fields

The fields are also spatially at right angles, so the magnitude of their resultant is given, by Pythagoras' theorem, as

$$|e_R| = \sqrt{e_1^2 + e_2^2} = E\sqrt{\cos^2 \omega t + \sin^2 \omega t} = E$$

The angle θ in Fig. 6.14 is given by

$$\tan^{-1} \frac{e_2}{e_1} = \tan^{-1} \frac{\sin \omega t}{\cos \omega t} = \omega t$$

The electric field stays the same magnitude, but rotates; one revolution every cycle. The magnetic field does the same – always at right angles to the electric field. Moving in the direction of propagation, in each successive plane the fields stay a constant magnitude and rotate: the rotating fields, at a given instant of time, are progressively delayed in angle along the direction of propagation. If one of the two waves were reversed in phase, the direction of rotation would be reversed.

This composite wave is called a circularly polarized wave: if, looking in the direction of propagation, the fields rotate clockwise, we call it *right hand circular polarization*; if anticlockwise, *left hand circular polarization*.

It is possible to design an antenna (or primary feed) which generates directly a circularly polarized wave: this antenna will also receive circularly polarized radiation more efficiently than a dipole. A popular design is in the form of a helical rod in front of a ground plane (*see* Fig. 6.15).

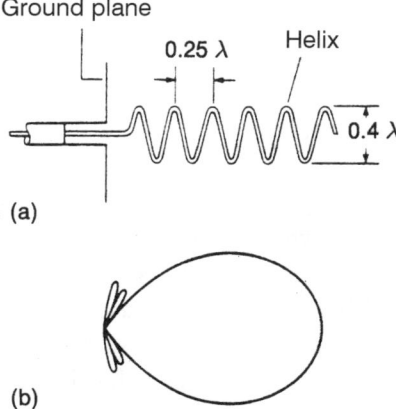

Fig. 6.15 (a) a helical antenna, (b) its polar diagram

An antenna for circular polarization must be designed for either left hand or right hand polarization – the two are orthogonal and thus can, in principle, be involved in frequency reuse.

An example of a satellite link

One application of satellite transmission that will almost certainly continue to be important is communication with mobile receivers. We shall take as an example communication to and from ships at sea, using a satellite in a geostationary orbit.

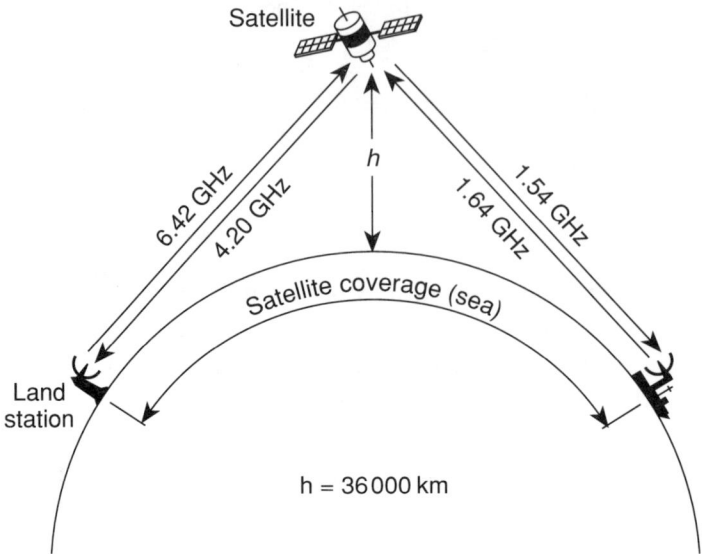

Fig. 6.16   A satellite link to ships at sea

Figure 6.16 shows the situation (not to scale), and frequencies typically used. The satellite stays in the same place in the sky, so transmissions to and from the fixed land station are in one direction, whereas the ships can be anywhere in the area of the satellite's coverage.

The fixed land station will have an antenna with a large aperture while those on the satellite will be much smaller, particularly that transmitting to the ships, since it has to have a wide beam. Assume that the ship's receiver has a dish antenna that can be steered to point at the satellite so that its beam need not be too wide and thus its gain too small.

Some typical data are shown in Table 6.1. Notice that the sign of the G/T ratio in decibels is positive for the land station receiver and negative for the ship station and satellite receivers. This is simply because in the first case the gain ratio is numerically greater than the equivalent noise temperature, whereas in the other cases it is less. The sign of the decibel value is significant and must be used correctly.

In the case of path loss and absorption the decibel values are often quoted (as here) with positive values, although, since a multiplier less than unity is represented, this value must be subtracted in a link budget calculation. The ambiguity arises because conventionally the *loss* of any network, in decibels, is always quoted as minus its *gain* in decibels, so that the inclusion of the word loss implies a reversal of sign.

The gain of the land station antenna at its receiving frequency works out to approximately 2.2×10^5. 32 dBK for G/T is equivalent to a ratio of 1585, yielding a value for T_e of about 140 K. Such a low value can be achieved for two reasons. First the antenna is looking out into space, so it sees a T_s of only a

Table 6.1

Fixed land station: parabolic dish antenna of diameter 13 m, used for both transmitting and receiving		
Transmitter	(6.42 GHz)	EIRP 60.0 dBW
Receiver	(4.20 GHz)	G/T 32.0 dBK

Ship station: steerable parabolic antenna of diameter 1.2 m, used for both transmitting and receiving		
Transmitter	(1.64 GHz)	EIRP 36.0 dBW
Receiver	(1.54 GHz)	G/T −3.5 dBK

Satellite		
Transmitter	(4.20 GHz)	EIRP −2.5 dBW
Transmitter	(1.54 GHz)	EIRP 18.0 dBW
Receiver	(6.42 GHz)	G/T −17.0 dBK
Receiver	(1.64 GHz)	G/T −13.2 dBK

Free-space loss between land station and satellite	
at 6.42 GHz	200.9 dB
at 4.20 GHz	197.2 dB

Free-space loss between ship and satellite	
at 1.64 GHz	188.9 dB
at 1.54 GHz	188.4 dB

Atmospheric absorption	
6.42 and 4.2 GHz	0.4 dB
1.64 and 1.54 GHz	0.2 dB

few degrees Kelvin, second the receiver must use a special low noise first-stage amplifier – probably a type known as a parametric amplifier. If the amplifier were cooled with liquid helium it would be possible to get T_e down as low as 20 or 30 K.

Because the land station receiver has such a high G/T value, the transmitter on the satellite transmitting to the land station can have low power, thus conserving satellite power: this accounts for its low EIRP value.

Satellite link budgets

Table 6.2 lists link budgets for the land station to satellite and satellite to ship station links. Each decibel value is given to one decimal place, and is given the correct sign so that the process is formally one of addition.

The definition of G/T does not include any noise originating from the transmitter, so we now consider noise, at the receiver terminals, which comes from the transmitter. Denote the spectral power density of this noise at the receiver terminals as N_T. Remembering that the carrier power at the receiver terminals is C, since N_T and C must both have been reduced by the same factor in the journey from the transmitter, the carrier to noise density ratio at the transmitter must have been C/N_T.

Table 6.2

Land station to satellite	
EIRP	60.0 dBW
Atmospheric absorption	−0.4 dB
Free-space path loss	−200.9 dB
$1/k$	228.6 dB
G/T	−17.0 dBK
$(C/N_0)_u$	70.3 dBHz
Satellite to ship	
EIRP	18.0 dBW
Atmospheric absorption	−0.2 dB
Free-space path loss	−188.4 dB
$1/k$	228.6 dB
G/T	−3.5 dBK
$(C/N_0)_d$	54.5 dBHz

Taking account of this transmitter noise, the carrier to noise density ratio at the receiver becomes[6]

$$\frac{C}{N_T + N_0}$$

If we turn these ratios over and deal with noise density to carrier ratios it will be seen that

$$\frac{\text{total noise density}}{\text{carrier power}} = \frac{N_T}{C} + \frac{N_0}{C}$$

This argument applied more generally leads to the conclusion that the overall noise to carrier ratio (inverse of carrier to noise ratio) for a system can be obtained by adding together the noise to carrier ratios for the different sources of noise, calculating each separate ratio at whatever point in the system is convenient.

In the case of a satellite we have an uplink and a downlink. If we assume that the transmitter noise in the land station transmitter is negligible compared to the transmitted carrier power, then the carrier to noise density ratio at the input terminals of the satellite transponder is $(C/N_0)_u$ as shown in the link budget (Table 6.2). Since noise generated in the satellite receiver has been referred to its input (in establishing the receiver gain/noise temperature ratio), $(C/N_0)_u$ is also the carrier to noise density ratio at the transmitter, for the downlink (assuming that the transmitter itself produces very little noise compared to its signal power). The carrier to noise density ratio produced purely by the downlink is that shown in the link budget as $(C/N_0)_d$. The carrier to noise density ratio for both links combined is obtained by inverting both, adding and reinverting the sum.

The overall carrier to noise density ratio for both up- and downlink is as follows:

---

[6] The total noise power is taken to be the sum of the contributing noise powers because the noise sources are assumed to be uncorrelated.

since

$(C/N_0)_u = 70.3\,\text{dBHz}$

$(N_0/C)_u = -70.3\,\text{dBHz} = 9.3 \times 10^{-8}$

similarly

$(N_0/C)_d = -54.5\,\text{dBHz} = 3.55 \times 10^{-6}$

so, overall

$$N_0/C = (N_0/C)_u + (N_0/C)_d$$
$$= 3.643 \times 10^{-6} = -54.4\,\text{dBHz}$$

$C/N_0 = 54.4\,\text{dBHz}$

In this case, noise in the uplink has very little effect; the channel is said to be down-link limited.

   If there is interference then this can be dealt with in a similar way: the ratio of the power of the interference divided by the bandwidth to the carrier power at the point at which the interference is deemed to enter the system can be added in to produce an overall noise-plus-interference density to carrier ratio which is then inverted.

   The satellite transmitter power for transmission to the ships will probably be around 200 W, but the distance is large and the beam is wide. Also note, the values given for atmospheric absorption do not include any margin for heavy rain. Comparing the resulting carrier to noise ratio with that for the terrestrial microwave link one will see that signalling rates must be much smaller even allowing for the fact that the data is coded in such a way that lower signal to noise ratios at the demodulator are acceptable.

6.4 CALCULATIONS

Using the data from Table 6.1:

6.1 Set out the link budgets for the ship to land station direction and hence calculate the overall carrier to noise density ratio for the channel in that direction.

6.2 Calculate the angular beamwidths for the land station antenna at both its transmitting and receiving frequencies and thus determine how accurately it must point at the satellite.

6.3 Deduce the level of transmitter power used by the land station.

6.4 Calculate the ship transmitter power and receiving system noise temperature.

7 Further insights using the Smith chart

Although the Smith chart was invented as an aid to calculations relating to line transmission, it has been found that using it gives very powerful insights into the behaviour of all types of wave-guiding systems involving reflections.

7.1 THE SMITH CHART

A Smith chart is a type of graph, in polar form, from which it is possible to deduce the input impedance of a length of terminated line, assuming that the line has negligible attenuation over the length concerned. One needs to know the characteristic impedance of the line (assumed to be resistive), the wavelength on the line and either the standing wave ratio or the return loss. One also needs to know the position of a voltage maximum or minimum. A Smith chart graticule is shown in Fig. 7.1.

Pads of Smith chart forms are available from technical stationers. These, being about twice the diameter of the charts printed in this book, can be read with rather more accuracy than is possible here.

Theory of the Smith chart

At any point on a transmission line, not terminated by its characteristic impedance, which is carrying a sinusoidal wave, there will be a voltage due to the wave travelling towards the load, that can be represented by the phasor V_i, and a voltage due to the reflected wave returning from the load, V_r. Calling the distance to the load L, we can define a reflection coefficient at this point

$$\rho_L = \frac{V_r}{V_i}$$

As the ratio of two phasors, ρ_L will be a complex quantity with both magnitude and angle. Because we have assumed negligible attenuation, the magnitude of ρ_L will not change – we shall call it ρ – but the angle will depend on L and on the nature of the terminating impedance Z_T.

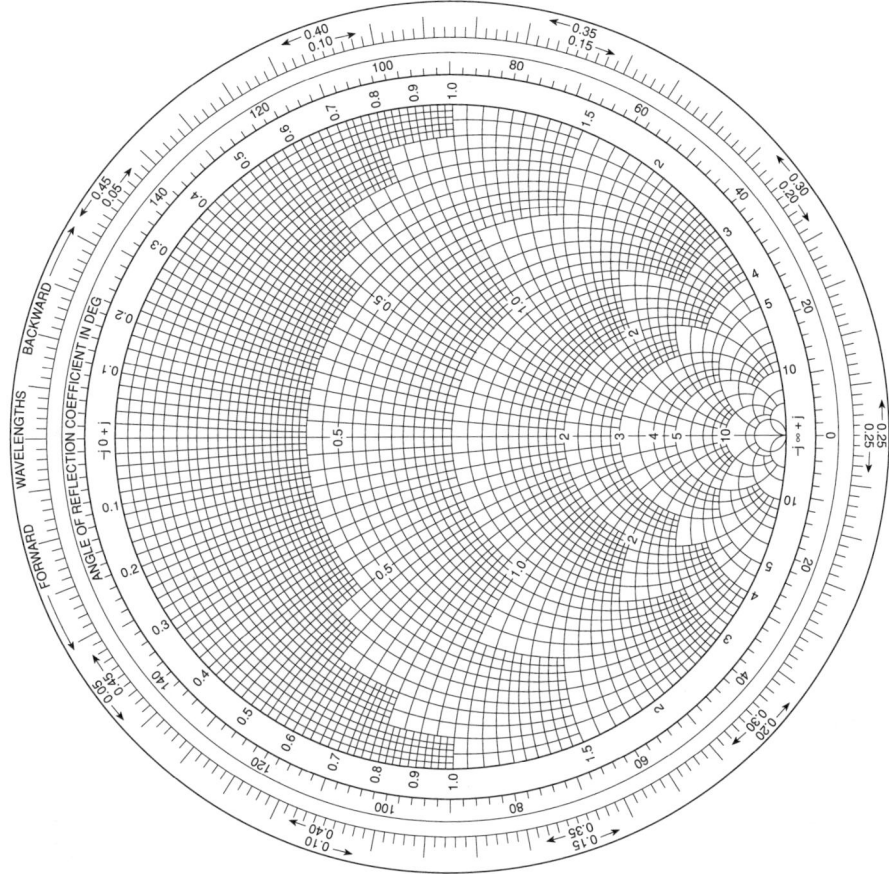

Fig. 7.1

Look at Fig. 7.2 which shows the incident and reflected voltages and currents on the line at distance L from the load – the relative directions of voltages and currents in the two waves are determined by the directions of energy flow.

$$Z_{in} = \frac{V_i + V_r}{I_i - I_r} = \frac{V_i}{I_i} \frac{1 + V_r/V_i}{1 - I_r/I_i}$$

and since V_i/I_i and V_r/I_r both equal Z_0,

$$\frac{V_r}{V_i} = \frac{I_r}{I_i}$$

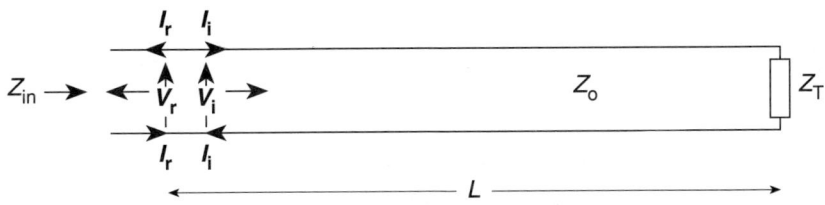

Fig. 7.2 Incident and reflected voltages and currents on a transmission line

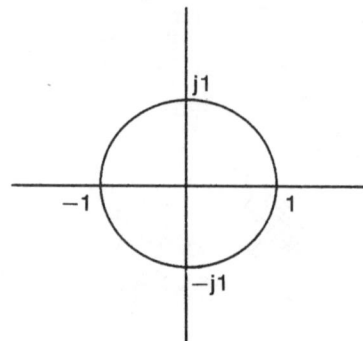

Fig. 7.3　The unit circle on an Argand diagram

so I_i/I_r also equals ρ_L and

$$Z_{in} = Z_0 \frac{1 + \rho_L}{1 - \rho_L}$$

Z_0 is resistive at the frequencies for which the Smith chart can be used, and so it is possible to work in terms of a *normalized input impedance* z_{in}, which is defined as Z_{in}/Z_0 and which has the same angle (same ratio of real to imaginary parts) as Z_{in}. This allows one to use the same theory for a transmission line of any characteristic impedance, and in these terms

$$z_{in} = \frac{1 + \rho_L}{1 - \rho_L} \tag{7.1}$$

ρ_L can have any magnitude between 0 and 1 and any angle, so that on an Argand diagram it must lie within the unit circle, that is, a circle centred on the origin and passing through the points 1, j1, −1 and −j1 (*see* Fig. 7.3).

Now write $\rho_L = u + jv$ and $z_{in} = r + jx$ thus

$$r + jx = \frac{1 + u + jv}{1 - u - jv}$$

Rationalizing the denominator

$$r + jx = \frac{(1 + u + jv)(1 - u + jv)}{(1 - u)^2 + v^2} = \frac{1 - u^2 - v^2 + j2v}{(1 - u)^2 + v^2}$$

Equating real and imaginary parts

$$r = \frac{1 - u^2 - v^2}{(1 - u)^2 + v^2} \tag{7.2}$$

and

$$x = \frac{2v}{(1 - u)^2 + v^2} \tag{7.3}$$

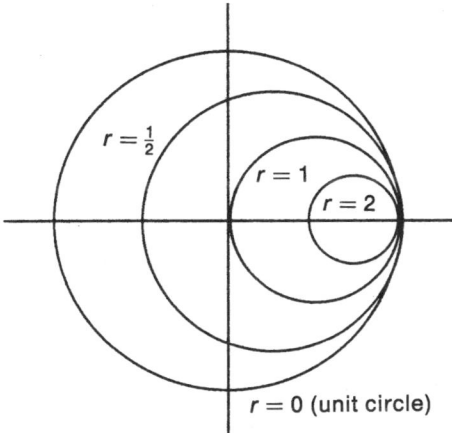

Fig. 7.4   Circles of constant normalized resistance

Equation (7.2) can, with some difficulty, be manipulated into the form

$$\left(u - \frac{r}{1+r}\right)^2 + v^2 = \frac{1}{(1+r)^2} \tag{7.4}$$

This can be checked by expanding out Equations (7.2) and (7.4) and showing that they give the same terms.

Equation (7.4) is the equation of a circle on the Argand diagram centred at the point $u = r/(1+r)$, $v = 0$ and of radius $1/(1+r)$. If the centre of the circle is at $r/(1+r)$ and the radius is $1/(1+r)$, the circle must cut the real axis at the point $r/(1+r) + 1/(1+r) = 1$.

If $r = 0$, the centre is at zero and the radius is 1. Figure 7.4 shows circles on the Argand diagram for $r = 0$, $r = \frac{1}{2}$, $r = 1$ and $r = 2$.

Equation (7.3) can be manipulated into

$$(u - 1)^2 + \left(v - \frac{1}{x}\right)^2 = \frac{1}{x^2} \tag{7.5}$$

Equation (7.5), for different values of x, represents a set of circles centred on points distant $1/x$ from the point $1 = j0$ on the line AB, and of radius $1/x$, in Fig. 7.5. Bearing in mind that values of v and u are only valid within the unit circle, Fig. 7.5 shows curves of constant x (arcs of circles) on the Argand diagram.

By superimposing Figs. 7.4 and 7.5 one should see that the Smith chart is in fact the unit circle of an Argand diagram with circles of constant normalized resistance and arcs of constant normalized reactance drawn upon it.

A circle representing all possible complex values of the reflection coefficient can be drawn on the Smith chart. Figure 7.6 shows the reflection coefficient circle for $\rho = 0.5$ (one quarter of the energy reflected back). The radius of the circle is half the radius of the edge of the Smith chart, i.e. half the distance from the origin to

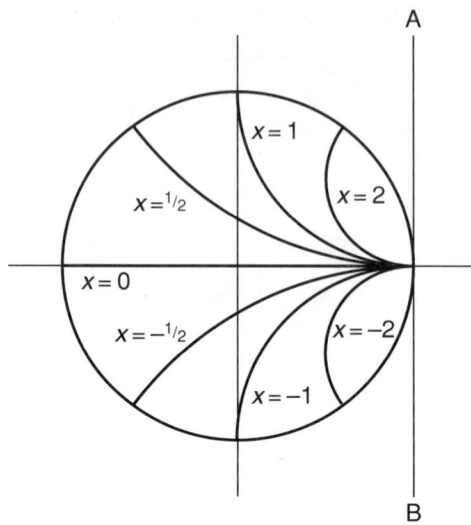

Fig. 7.5   Arcs of constant normalized reactance

the unit circle. The voltage standing wave ratio corresponding to this reflection coefficient magnitude is given by the formula

$$S = \frac{1 + \rho}{1 - \rho} = \frac{1.5}{0.5} = 3$$

Since Equation (7.1) gives

$$z_{\text{in}} = \frac{1 + \rho_{\text{L}}}{1 - \rho_{\text{L}}}$$

it follows that the value of S is the same as the value of z_{in} when ρ_{L} is real and positive, so the reflection coefficient circle goes through the point $r = 3$ (marked S) on the real axis. If the information available is the value of the VSWR, this allows one to draw the reflection coefficient circle without calculating the value of ρ; in fact, the reflection coefficient circle is often referred to as a *standing wave circle*.

As one moves from one point on a line to another the angle of the reflection coefficient changes. Remembering that there is an increasing phase lag with distance in the direction of propagation, as one moves towards the load the phase of the incident wave voltage decreases while that of the reflected wave voltage increases. Suppose a small change in position towards the load causes the incident wave voltage phase to decrease by θ and that of the reflected wave to increase by θ. The change in angle of $V_{\text{r}}/V_{\text{i}}$ is $\theta - (-\theta) = 2\theta$, so, the angle of the reflection coefficient increases as you move towards the load at twice the rate at which the phase of the travelling wave changes. A complete $360°$ cycle of reflection coefficient angle occurs in half a wavelength on the line.

If a line is drawn through the centre of the Smith chart, the complex value of the reflection coefficient at the point at which it crosses the reflection coefficient circle at one side is minus its value where it crosses the circle at the other side (remember that this is an Argand diagram upon which complex numbers are plotted). Starting from z_{in}, where the complex reflection coefficient is ρ_L, crossing the diagram in the way indicated gives a reading on the graticule which is

$$\frac{1+(-\rho_L)}{1-(-\rho_L)} = \frac{1-\rho_L}{1+\rho_L}$$

Comparing this with Equation 7.1 shows that the reading gives the inverse of z_{in}, i.e. the normalized input admittance. For example, one point on the $\rho = 0.5$ circle on Fig. 7.6 is the point P where from the graph we read that

$$z_{in} = 0.78 + j1$$

The normalized input admittance is

$$y_{in} = \frac{1}{0.78 + j1}$$

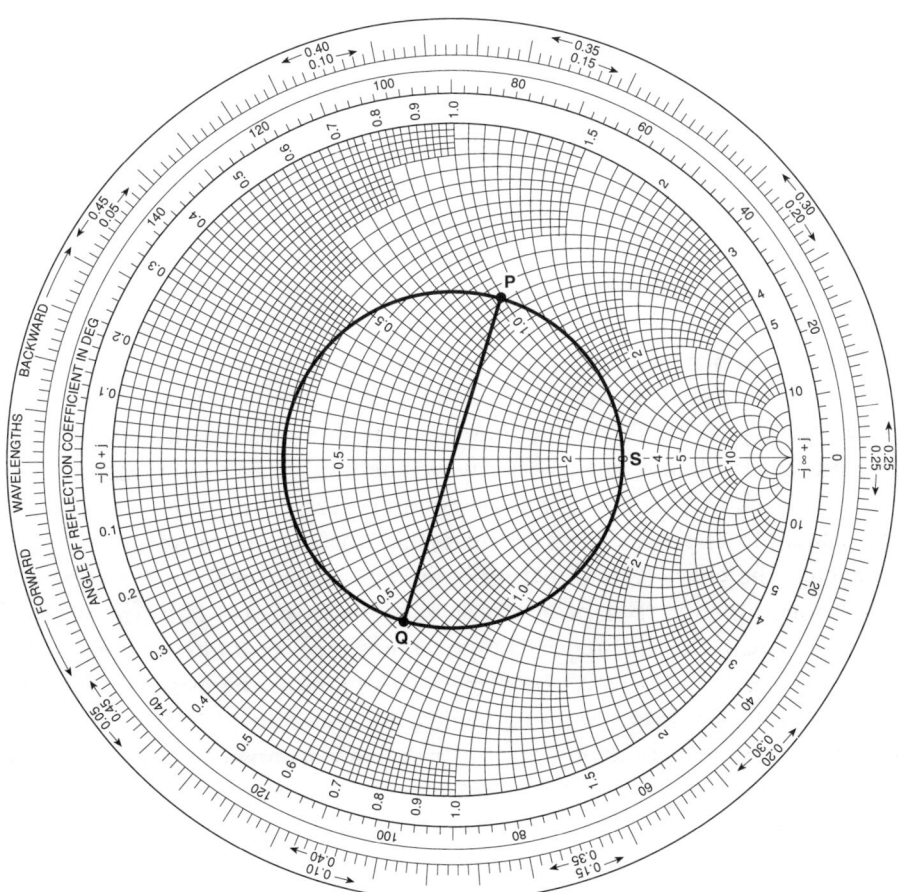

Fig. 7.6

Rationalizing the denominator

$$y_{in} = \frac{0.78 - j1}{0.78^2 + 1}$$

which works out to

$$y_{in} = 0.48 - j0.62$$

Crossing the Smith chart from P to the point Q, we find this value at Q.

If a line is matched, then the normalized input impedance (and admittance) is $1 + j0$, i.e. the centre of the chart which for this reason is sometimes known as the *matching point*.

Because the Smith chart is used at frequencies for which the losses per wavelength are negligible, the locus of reflection coefficient is taken to be a circle, however, plotting the reflection coefficient on the chart over many wavelengths would produce a gradual spiral in toward the centre.

Although the theory of the Smith chart has been developed by reference to transmission line, it is also applicable to waveguide. The impedances involved will then be normalized wave impedances.

7.2   RESONANT LINE SECTIONS

It is easy to see from the Smith chart how a resonant section of line, either open-circuited or short-circuited, works. Most often a short-circuited line section will be used because it is easier to ensure that a line is truly short-circuited than open-circuited, and in the case of twin-wire line the short-circuited section is likely to be more mechanically rigid. A short-circuited section of line is often referred to as a *stub*.

The next argument will be much easier to follow if we use some numbers, so assume that the design wavelength on the line is 10 m – this corresponds to a signal frequency of 30 MHz if the line wires are in air so that the phase velocity is c – and that the line is one-quarter wavelength, i.e. 2.5 m long, short-circuited at the end. Look at Fig. 7.7. The short circuit corresponds to the point A on the Smith chart. Moving away from the load (the short circuit) means moving clockwise around the chart (this direction is marked 'backward' on the edge of the chart) through one-quarter of a wavelength, which means half-way round the chart, to B, where the (normalized) input impedance is infinite.

In practice there will always be some losses (conductor resistance and radiation): suppose these losses were equivalent to a standing wave ratio of 10 on the stub. The input would now have a normalized impedance of 10 – i.e. the input impedance would be $10R_0$ – shown at point C on the chart, and the load would appear to be $R_0/10$, at D, rather than a short circuit (although in practice the losses would be distributed).

How does the stub respond to a lower frequency? A lower frequency means a longer wavelength, so the stub is now less than a quarter wavelength long. We are

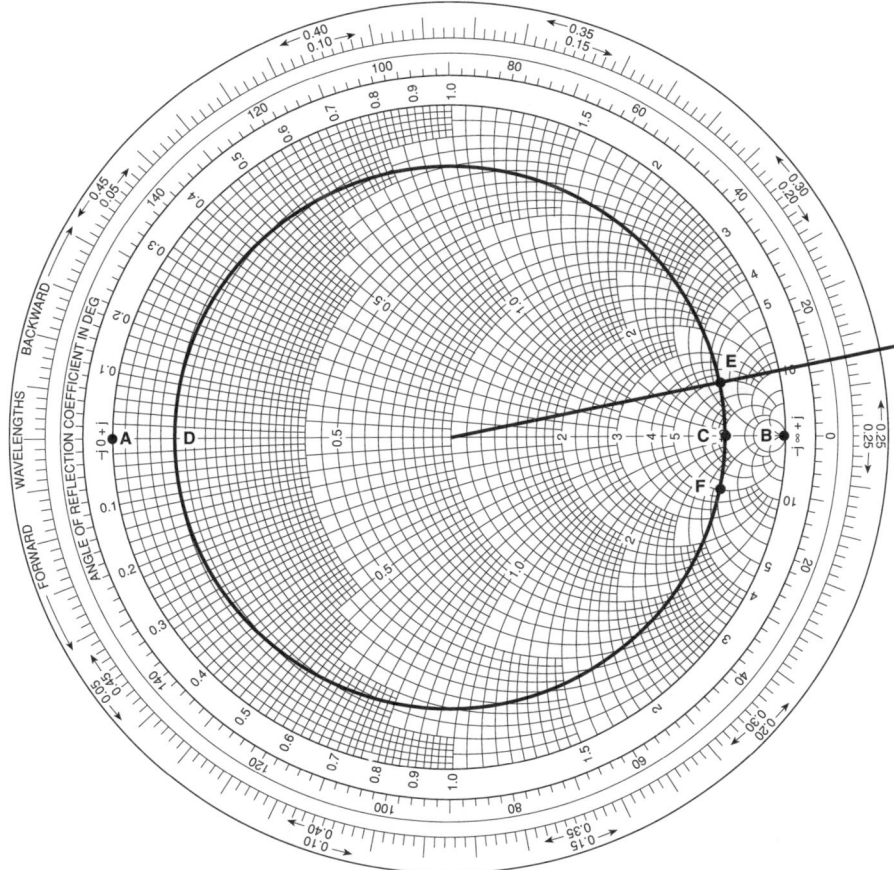

Fig. 7.7

interested in the wavelength that brings the input impedance to the point marked E on the chart: this is significant because at this point the input impedance has an inductive reactance equal in magnitude to its resistance, its normalized input impedance being $5 + j5$. Similarly, by raising the frequency and thus making the wavelength shorter the stub will be longer than one-quarter wavelength, and, at a particular frequency its normalized input impedance will be $5 - j5$, at the point F. Over the frequency range considered, then, the stub acts like a parallel tuned circuit with 3 dB frequencies that can be determined from the positions of E and F on the Smith chart.

Take point E. From the calibrations on the edge of the chart we read that the distance from the short-circuit is 0.234λ. Hence λ is $2.5/0.234\,\text{m} = 10.68\,\text{m}$.

Assuming a phase velocity of 3×10^8 m/s, the frequency is 28.08 MHz.

By similar reasoning, the frequency associated with the point F is 31.92 MHz. The bandwidth is 3.84 MHz.

And so the Q, given by resonant frequency/bandwidth, is $30/3.84 = 7.8$.

It is seen that when a resonant line section is used as a tuned circuit its Q depends on the losses.

The balance to unbalance transformer (Balun)

When a coaxial cable is used to feed an antenna, an output balanced about earth is usually required. The outer of the coax is generally at earth potential, at least at the transmitter, but a balanced output can be obtained by attaching an extra conducting sleeve, one-quarter wavelength long, as shown in cross-section in Fig. 7.8. The extra outer forms with the original outer a coaxial line which, being short-circuited at A, presents an open circuit at B. The sleeve takes on earth potential, but the potential of the original outer is not tied to earth at B and so the alternating voltage between the inner and the original outer at B balances itself about earth potential. The space between the sleeve and the coax outer will be resonant and carry a standing wave.

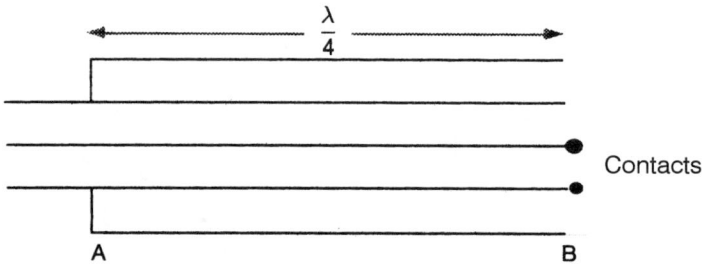

Fig. 7.8   A Balun

7.3   A QUARTER-WAVE TRANSFORMER

Figure 7.9 shows a transmission line of characteristic impedance Z_0 feeding a load Z_L via a quarter-wave section of line of characteristic impedance Z_m. The load impedance normalized to the quarter-wave section is Z_L/Z_m. One-quarter wavelength away (half-way round a Smith chart) the normalized impedance seen by the main line is Z_m/Z_L, so the actual impedance is $(Z_m/Z_L) \times Z_m$.

Hence the matching condition is

$$Z_0 = Z_m^2/Z_L \qquad \text{i.e.} \qquad Z_m = \sqrt{Z_0 Z_L}$$

Since Z_0 and Z_m will both be resistive (at the frequencies at which Smith charts can be used, characteristic impedances are resistive) it follows that Z_L must be resistive for a quarter-wave match to be possible.

Now suppose we want to know how far off frequency we can go and still have a reasonable match.

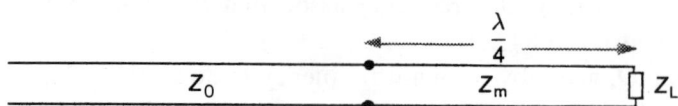

Fig. 7.9   A quarter-wave impedance transformer

Again, it will be easiest to take some values. Suppose the load is 112.5 Ω and the characteristic impedance of the line is 50 Ω. The design frequency is 300 MHz. We want to find the extent of the mismatch if the frequency can vary by ±10 per cent.

The quarter-wave section will have a length of

$$\frac{c}{4f} = \frac{3 \times 10^8}{4 \times (3 \times 10^8)} \text{ m} = 25 \text{ cm}$$

The required matching section impedance is

$$Z_m = \sqrt{50 \times 112.5} = 75 \, \Omega$$

Let us see first what is the mismatch without the matching section.

The normalized load impedance, and also the VSWR is

$$112.5/50 = 2.25$$

The standing wave circle is plotted on Fig. 7.10.

The reflection coefficient is $1.25/3.25 = 0.385$ which indicates a fraction of energy reflected $0.385^2 = 0.148$, i.e. nearly 15 per cent.

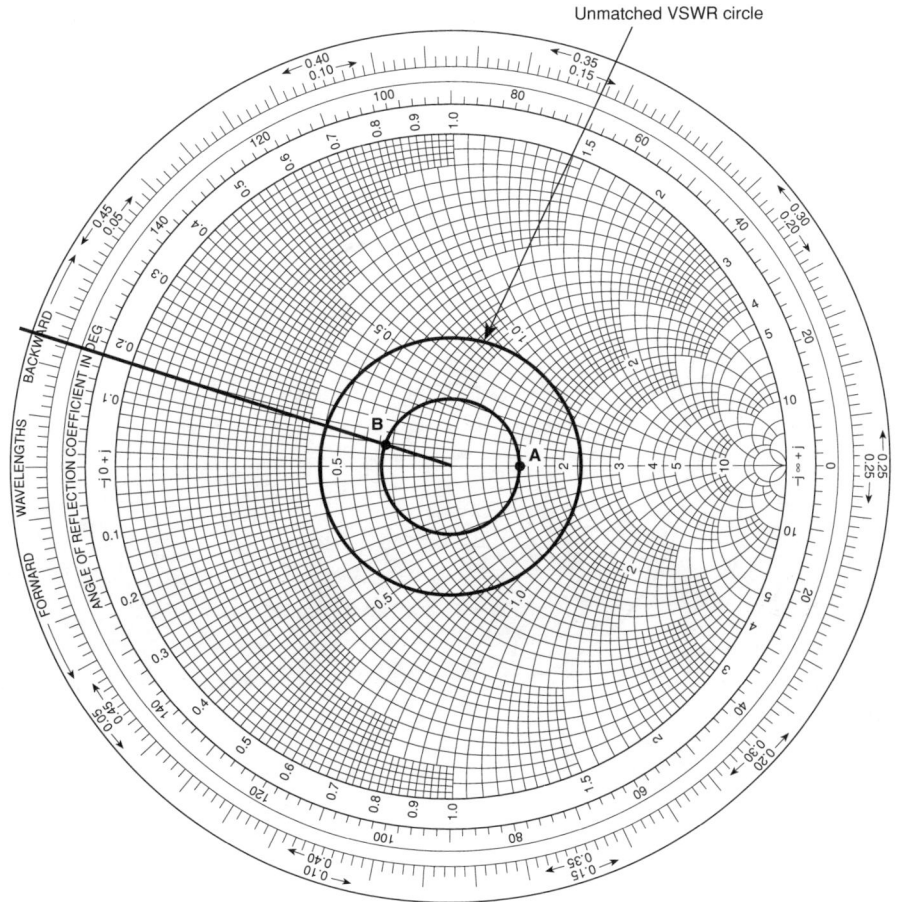

Fig. 7.10

At the design frequency, the quarter-wave section matches the load: what is the mismatch at 330 MHz?

At this frequency the wavelength is $(3 \times 10^8)/(3.3 \times 10^8) = 90.9$ cm, so 25 cm $= 0.275\lambda$.

The load impedance normalized to the matching section is $112.5/75 = 1.5$. The standing wave circle is shown on Fig. 7.10.

Starting at the point marked A, we move round the chart, in a clockwise direction (away from the load) until we reach the point B. Now, however, we are in some trouble because the graticule is difficult to read accurately with such a small standing wave circle, so we move on to Fig. 7.11. This is an expanded version of the centre of the Smith chart, out to a VSWR of 2, and the grids for this are again available from technical stationers. The normalized impedance at B can now be read as

$$0.67 + j0.09$$

So, the load presented to the main line is

$$75(0.67 + j0.09) = 50 + j6.8$$

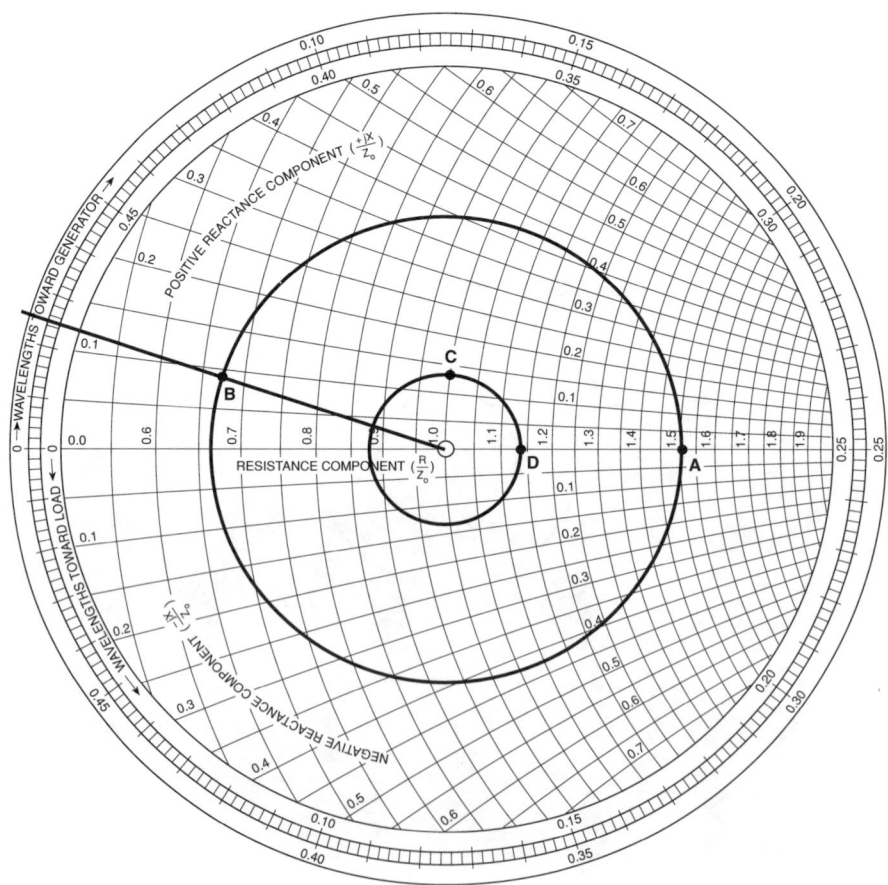

Fig. 7.11

If we now renormalize this to the main line impedance we can plot a standing wave circle representing the mismatch on the main line.

$$(50 + j6.8)/50 = 1 + j0.135$$

This is plotted on Fig. 7.11 and marked C. A standing wave circle drawn through C allows one to read off the VSWR on the main line, at D, giving $1.14:1$, a value that would normally be regarded as quite acceptable.

If we carry out the same procedure for a frequency of 270 MHz we shall get the same result.

Broad-banding

It is possible to make the match broader band by using more than one quarter-wave step. Let us continue with the same example and now suppose that we want to use the line over a frequency range from 200 to 400 MHz. We still need to design for 300 MHz since this is the mid-band frequency. Using the same method as previously, for a single quarter-wave section of characteristic impedance $75\,\Omega$, the VSWR at 200 and at 400 MHz is just over $1.5:1$.

Now suppose we insert two quarter-wave sections between the main line and the load. What must their characteristic impedances be? Look at Fig. 7.12. The load at A, normalized to Z_2 gives Z_L/Z_2. At B, this transforms to Z_2/Z_L and denormalizes to Z_2^2/Z_L. Normalizing this to Z_1 gives Z_2^2/Z_LZ_1 which, at C, transforms to Z_LZ_1/Z_2^2 and denormalizes to $Z_LZ_1^2/Z_2^2$, which must equal Z_0.

$$\frac{Z_LZ_1^2}{Z_2^2} = Z_0 \qquad \frac{Z_2}{Z_1} = \sqrt{\frac{Z_L}{Z_0}}$$

This does not indicate unique values for Z_1 and Z_2. However, another way of looking at a quarter-wave transformer is to think of the mismatches at each end as generating two reflected waves which return towards the load in anti-phase and thus cancel each other out. Thinking in this way it seems sensible, in order to achieve a broad-band result, that the values of Z_1 and Z_2 should be chosen so as to equalize the reflection at each junction. Looking again at Fig. 7.12, the reflection coefficient at C if the first transformer section were matched would be

$$\frac{Z_1 - Z_0}{Z_1 + Z_0}$$

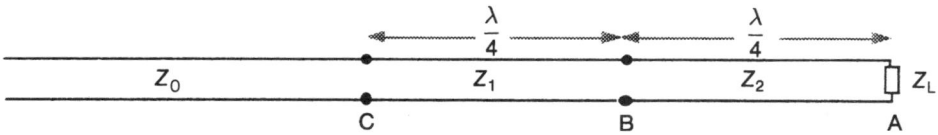

Fig. 7.12   A double quarter-wave impedance transformer

Taking the same consideration at B and at A leads to the requirement

$$\frac{Z_1 - Z_0}{Z_1 + Z_0} = \frac{Z_2 - Z_1}{Z_2 + Z_1} = \frac{Z_L - Z_2}{Z_L + Z_2}$$

From the first and last terms

$$\frac{1 - Z_0/Z_1}{1 + Z_0/Z_1} = \frac{1 - Z_2/Z_L}{1 + Z_2/Z_L}$$

A little algebra shows that this means

$$\frac{Z_0}{Z_1} = \frac{Z_2}{Z_L}$$

So, we require

$$\frac{Z_2}{Z_1} = \sqrt{\frac{Z_L}{Z_0}} \qquad \text{and} \qquad Z_2 Z_1 = Z_L Z_0$$

For the values in our example

$$\sqrt{Z_L/Z_0} = 1.5 \qquad \text{and} \qquad Z_L Z_0 = 5625$$

This yields $Z_1 = 61.24\,\Omega$ and $Z_2 = 91.86\,\Omega$, which we shall round to

$$Z_1 = 60\,\Omega \qquad \text{and} \qquad Z_2 = 90\,\Omega$$

Now consider how the match behaves at 200 MHz. The wavelength at this frequency is 1.5 m, so 25 cm is 0.167λ.

Looking again at Fig. 7.12, Z_L normalized to Z_2 is

$$112.5/90 = 1.25$$

A 1.25 VSWR circle has been drawn on Fig. 7.13. Starting at point A on the circle and moving away from the load 0.167λ brings us to point B. (The calibrations on the diagram are not very helpful here because they start from a zero on the left, so one has to count round.) We read the value at B to be

$$0.88 - j0.17$$

Notice that normalized reactance values in the bottom half of the diagram are negative. Denormalizing this from the 90 Ω section and renormalizing to 60 Ω gives

$$1.32 - j0.255$$

which is plotted at B'. Moving away from the load 0.167λ from B' brings us to point C which is

$$0.72 - j0.12$$

Denormalizing from 60 Ω and renormalizing to 50 Ω gives

$$0.86 - j0.14$$

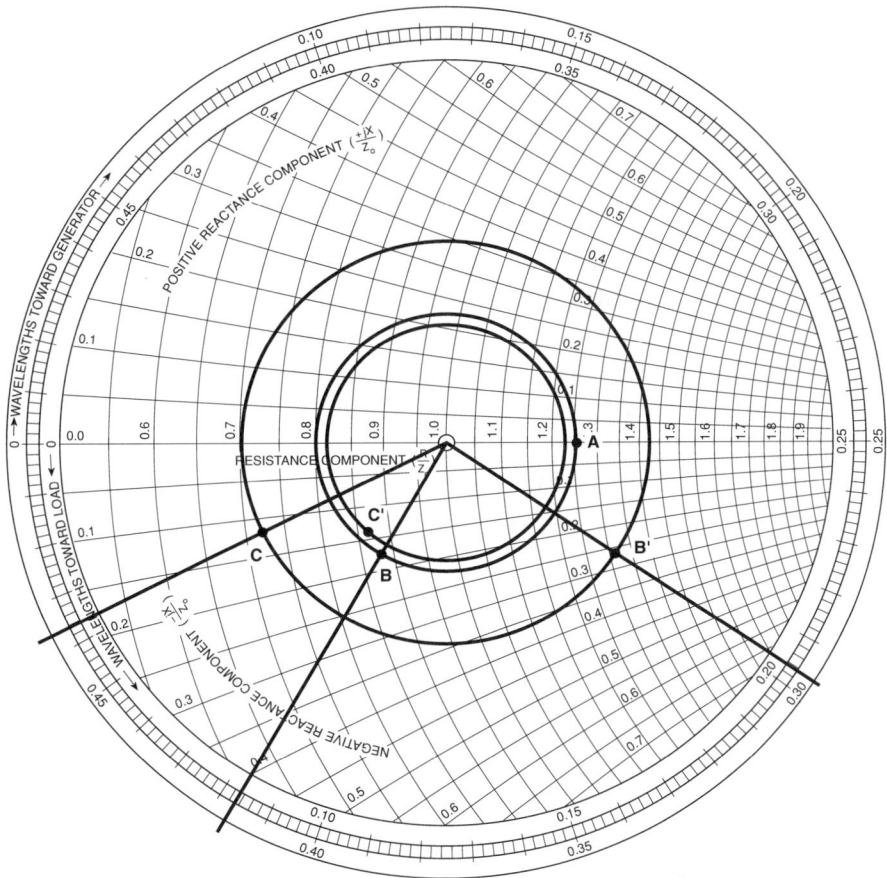

Fig. 7.13

which is plotted at C', and which, one will see, indicates a standing wave ratio of about 1.24 : 1.

You might like to check the match at a couple of frequencies nearer to 300 MHz to see that it is progressively better.

Three quarter-wave steps would give a better broad-band match still. Pursuing the idea that the match occurs due to out-of-phase reflections that cancel gives credibility to a method that is often used to achieve a broad-band match at high frequencies where the wavelength is short; that is, to effect a gradual change of the physical dimensions which control characteristic impedance over several wavelengths.

The quarter-wave transformer would seem to have the disadvantage that the load has to be resistive. This is not, however, a problem: at high frequencies the line usually terminates in some sort of structure of which the input terminals can be defined as a matter of convenience, so one can look at the standing wave ratio near to the load and define the load terminals as being the point on the line nearest to the load at which the input impedance is resistive. The line can then be cut at this point and the quarter-wave transformer inserted.

7.4   STUB MATCHING

In practice an impedance transformer has to be designed into the system when it is constructed. If one terminates a transmission line and then finds the mismatch unacceptable, a means must be found to improve the match without breaking the line: this can be done with a stub.

In this case one needs to work with admittances rather than impedances: for every point on the Smith chart representing a line's impedance its admittance is represented by the point diagonally opposite on the chart.

Whatever the VSWR, the standing wave circle will cross the unity normalized conductance circle at some point. To achieve a match, one must simply connect an admittance across the line which has a normalized value equal and opposite to that of the line at the point concerned.

In principle the match can be made knowing the admittance of the load and the phase velocity on the line, but in practice this is not the sort of information that is immediately available. Instead, measurements are made with a *standing wave indicator*. A standing wave indicator is a section (of rigid twin-wire line or coaxial cable or waveguide) in which a small probe is inserted into the electric field to sample its value. The probe can be moved along the direction of propagation over a sufficient length that several electric field maxima and minima can be observed. These devices are very carefully designed so that they interfere with the signal being measured as little as possible.

By moving the probe and measuring the voltage picked up from the electric field one can deduce the ratio of the electric field maximum to its minimum values – this will be the same as the VSWR. The position of the standing wave pattern is best located through voltage (electric field) minima rather than maxima because the positions of the minima in the pattern are more sharply defined. The wavelength on the line can be deduced from the separation of the voltage minima, and it is also necessary to measure the distance from a minimum to the load.

The following is an example of a match on a twin-wire feeder taking a signal from a 30 MHz transmitter to an antenna. The VSWR is measured to be 3.2 : 1. Two adjacent voltage minima are 5 metres apart. The nearest minimum to the antenna is 2 metres from the antenna terminals.

Look at the Smith chart in Fig. 7.14: the measured value of VSWR allows us to draw a standing wave circle through the point A. At a voltage minimum the admittance is maximum and resistive, so the point A also represents the admittance of the voltage minimum nearest to the source. There are two choices for making the match: move towards the load to point B or move away from the load to point C. In either case the distance moved is approximately 0.082λ. λ is twice the distance between two minima, i.e. 10 m, so the distance to move is 82 cm. A match nearer the load is neater, so we will go for point B. Here, the normalized input admittance is j1.25, so we need to connect in parallel with the line at this point an admittance $-$j1.25. One might think at this point that we need to know

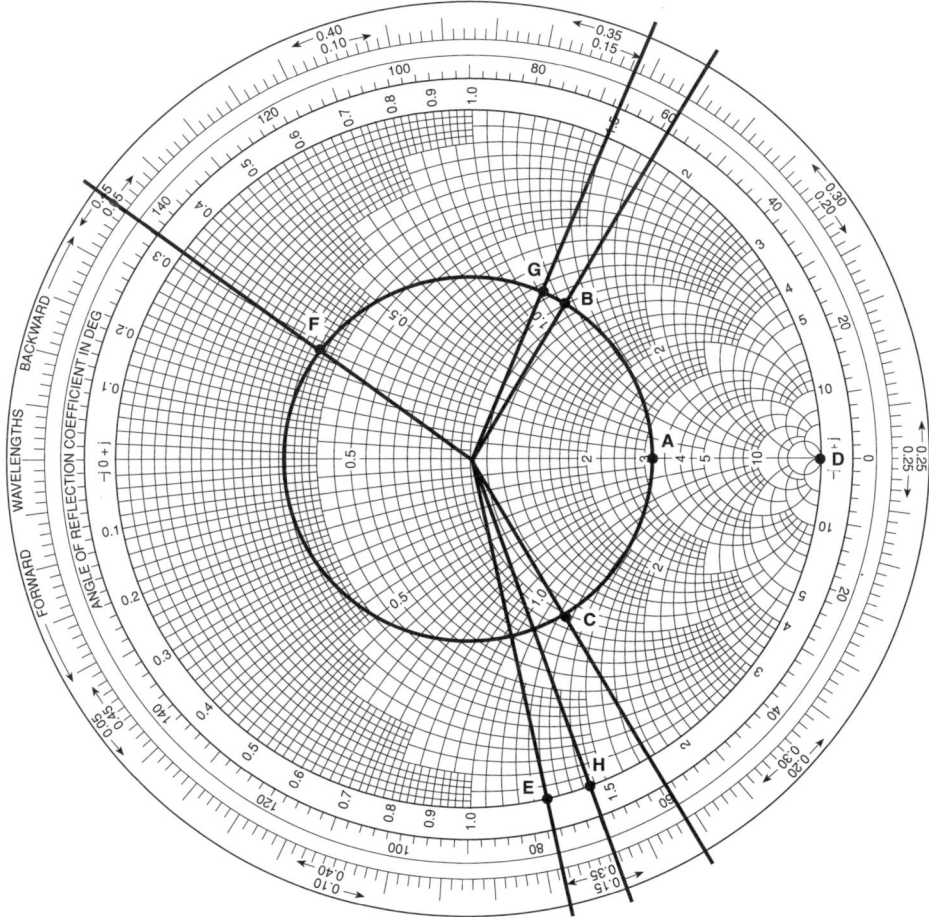

Fig. 7.14

the characteristic impedance of the line so as to calculate the actual admittance required, but we do not, because the admittance required is formed by using a short-circuited length of the same line.

At the short circuit the normalized admittance is infinite: this corresponds to the point D on Fig. 7.14. The standing wave circle (assuming no losses on the stub) is the circumference of the Smith chart, so, moving away from the short circuit to E gives a stub with an input admittance of $-j1.25$. D to E is 0.107λ, so the required stub length is 107 cm. If we had chosen to match at C, not only would the stub have been further from the load, but it would have had to be longer.

The matched line is shown in Fig. 7.15. There are large standing waves on the stub and between the stub and the load, but (if the match is perfect) no energy is returned to the transmitter. The energy is not dissipated (ideally), but as energy is fed in from the line the standing waves ensure that the total ratio of voltage to current at the load is such as to cause the load to accept energy at the same rate.

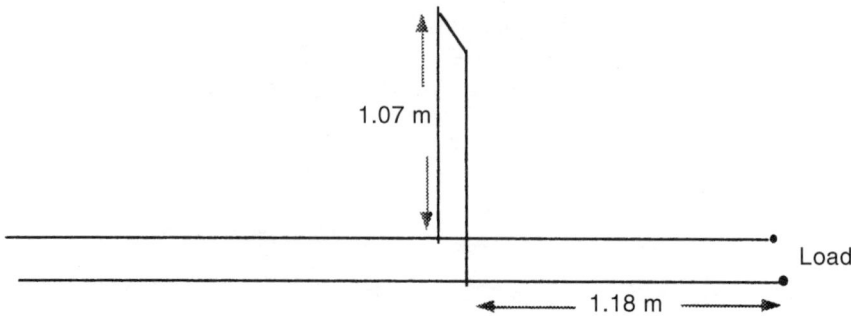

Fig. 7.15 A single stub match

How good is the match off frequency? Let us check it for a wavelength 10 per cent longer, i.e. 11 m. The load admittance has not changed; it is 0.2λ forward from the nearest minimum at the original wavelength – at point F on Fig. 7.14. The unmatched standing wave circle is not affected by the change of wavelength, but the matching point is now

$1.18/11 = 0.107\lambda$

backward from the load. This takes us to point G where we read the normalized admittance as

$0.85 + j1.14$

The length of the stub in wavelengths is now

$1.07/11 = 0.097\lambda$

This corresponds to point H on the Smith chart, which reads $-j1.42$ as the input admittance to the stub.

The load presented to the main line at the point of connection of the stub is

$0.85 + j1.14 - j1.42 = 0.85 - j0.28$

Putting this on the larger scale centre grid in Fig. 7.16 indicates a VSWR of about $1.4:1$ which is still very good compared to the unmatched VSWR of $3.2:1$.

Multiple stub matching

The method of matching just discussed is fine for an open wire feeder, but with a coaxial cable or a waveguide, fixing on a stub means major surgery. There is an alternative which is to use several stubs (at least two and frequently three), permanently installed, adjustable in length and separated by 3/8 of a wavelength. The lengths of the stubs are adjusted – usually by trial and error – until a match is obtained. To understand how this works look at Fig. 7.17 which is a Smith chart on which is drawn a circle equivalent to the circle of unit conductance but transposed $3\lambda/8$ towards the load. Figure 7.18 shows just two stubs. If the

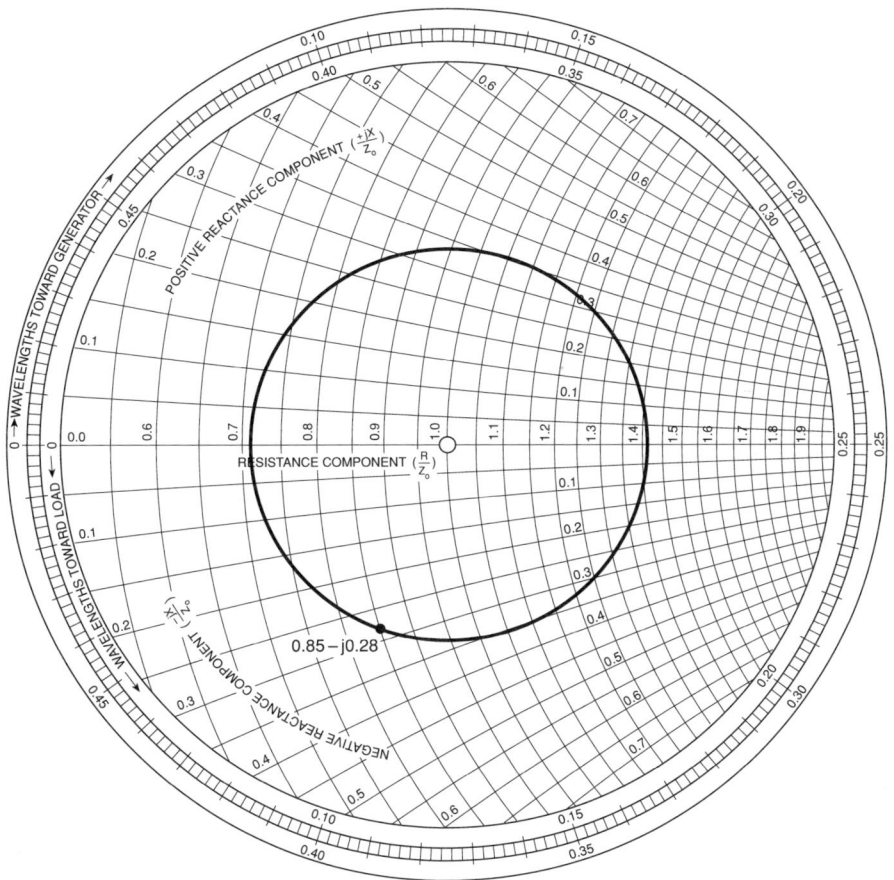

Fig. 7.16

real part of the normalized admittance at A looking towards the load has any value less than 2, then by using the stub at A to change the imaginary part, the normalized admittance can be made to lie on the displaced unit circle. Moving $3\lambda/8$ away from the load means that at B the normalized admittance lies on the true unit circle so that adjusting the length of the stub at B allows the line from there to the transmitter to be matched. For the small area of the chart, within the circle of normalized resistance 2, which cannot be matched, the second stub together with a third can give a match since we are $3\lambda/8$ further from the load and therefore out of the non-matching region.

Although, in principle, a match can be achieved with two stubs, frequently all three are adjusted: again an alternative view is that we are setting up cancelling reflections, and a useful principle, to give good performance away from the design frequency, is to try to spread the reflections equally between the stubs.

Lengths of waveguide with movable short circuits can be attached as stubs – to act as a shunt, the stub must be connected at the side wall – but most often this type of matching in a waveguide is achieved by inserting adjustable probes

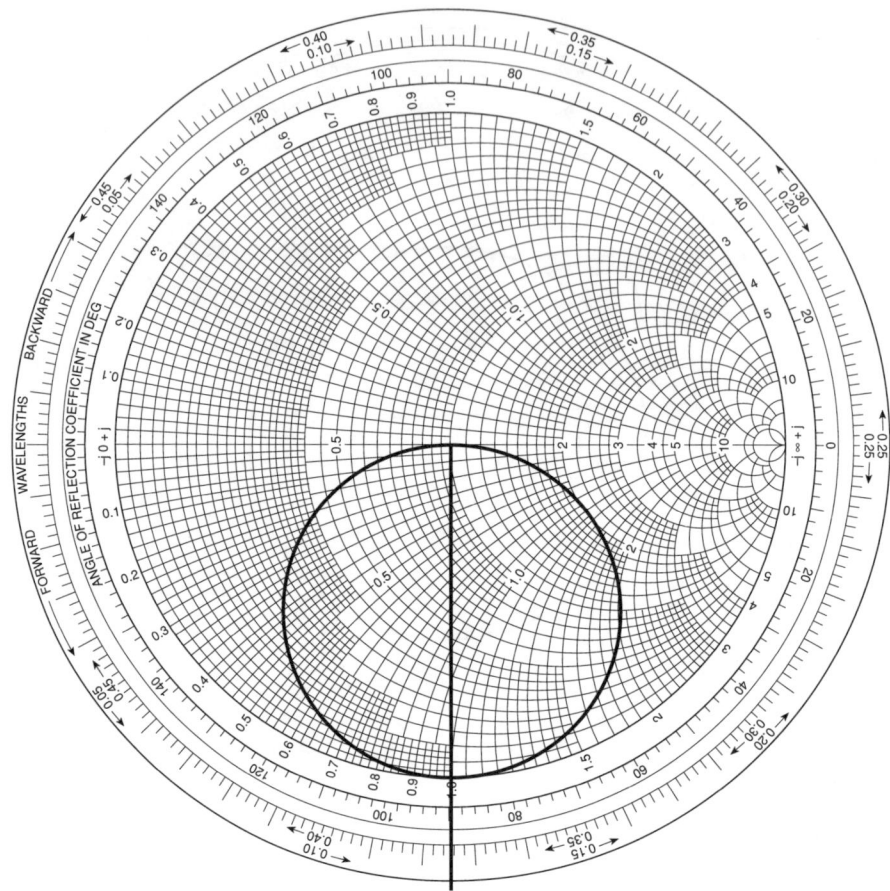

Fig. 7.17

in the top wall. In this case, clearly, the idea of cancelling reflections is the more useful.

7.5 GENERAL APPLICABILITY OF THE THEORY TO WAVEGUIDE

Design procedures at microwave frequencies tend more toward optics than circuit theory; nevertheless, many of the ideas of this chapter are germane. The wavelengths to be used are of course guide wavelengths.

Resonant cavities, which serve as tuned circuits for microwaves, can usefully be thought of as resonant lengths of waveguide of different cross-sections.

Quarter-wave transformers also find considerable use – a good example being to match the output of a microwave generator to a waveguide.

A waveguide horn is, in effect, a tapered match between the wave impedance in a waveguide and the wave impedance of free space.

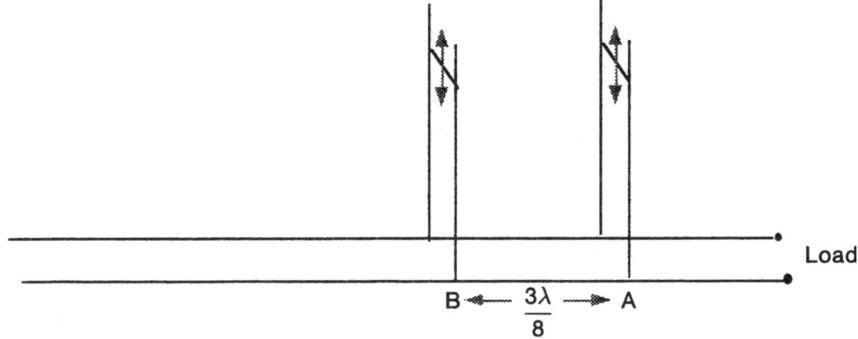

Fig. 7.18 A double stub matching system

Finally, as already indicated, triple-stub matching is routinely used in waveguide systems.

Direct applicability to optical fibre is less obvious – a resonant cavity in a laser can, perhaps, be thought of as a resonant section of optical waveguide, but these are generally many wavelengths long. There is no reason in principle why the ideas discussed should not be applied at optical frequencies in the future.

7.6 CALCULATIONS

Figure 7.19 is a Smith chart, with a standing wave circle drawn on it, relating to a mismatched transmission line of characteristic impedance $100\,\Omega$. The wavelength on the line is $15\,m$. The point marked A represents the normalized impedance of the load.

Carry out the required procedures on the Smith chart. Do not try to be more accurate than two significant figures.

7.1 What is the impedance of the load?

7.2 What is the normalized admittance of the load?

7.3 What is the voltage standing wave ratio on the line?

7.4 What is the distance from the load to the nearest voltage minimum in the standing wave pattern?

7.5 Deduce the position and length of a short-circuited stub of the same line required to match the line as near to the load as possible.

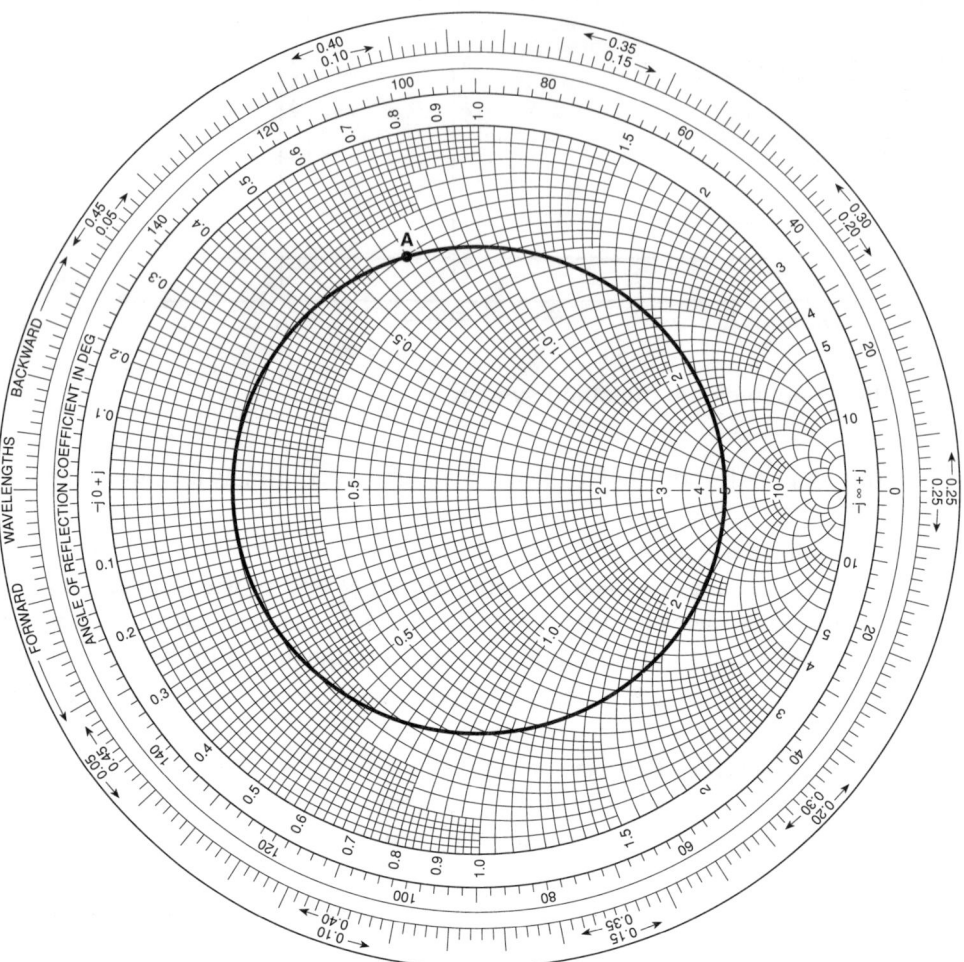

Fig. 7.19

Appendix 1
Complex exponential representations

A1.1   PHASOR REPRESENTATIONS

On an Argand diagram representing the complex plane (a plane on which can be plotted all the complex numbers) a line of length M joins the point $a + jb$ to the zero point – *see* Fig. A1.1.

The length $M = \sqrt{a^2 + b^2}$ and the angle on the diagram $\theta = \arctan(b/a)$. This is the normal basis for the representation of phasors such as V and I, and phasor operators such as Z and Y, each of which is an entity with a magnitude and an angle, by complex numbers. However, it can be seen from the diagram that

$$a = M \cos \theta \qquad \text{and} \qquad b = M \sin \theta \qquad \text{so} \qquad a + jb = M(\cos \theta + j \sin \theta)$$

and it can be shown mathematically, from the theory of infinite series, that

$$\cos \theta + j \sin \theta = e^{j\theta}$$

so the complex number $a + jb$ can also be written $M\,e^{j\theta}$. Hence a phasor or phasor operator, with a magnitude M and an angle θ, can be represented and manipulated mathematically either as $a + jb$, where $a = M \cos \theta$ and $b = M \sin \theta$, or as $M\,e^{j\theta}$.

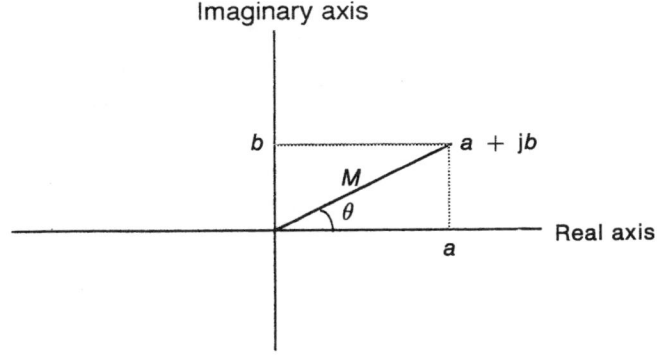

Fig. A1.1   The point $a + jb$ on an Argand diagram

A1.2   SINUSOID REPRESENTATIONS

In Fig. A1.2 a line is represented from the zero point on an Argand diagram, but this time the line is rotating anticlockwise with a constant angular frequency ω. The line is shown as having length A. Assuming that at time zero the line made an angle ϕ with the real axis, the angle made by the line with the real axis at time t is $(\omega t + \phi)$. The point at the end of the line can be represented as $A\,e^{j(\omega t + \phi)}$. A perpendicular dropped from the end of the rotating line to the imaginary axis meets it at the point P, and since the rotating line is centred on the zero point, the value of P is given by

$$\mathbf{P} = A\sin(\omega t + \phi)$$

Now at all times there is a direct relationship between the value, $A\,e^{j(\omega t + \phi)}$, of the point at the end of the rotating line and that of the point P, $A\sin(\omega t + \phi)$, and this relationship is maintained under all normal mathematical operations, so that the expression $A\,e^{j(\omega t + \phi)}$, which is easier to manipulate mathematically, can be substituted for $A\sin(\omega t + \phi)$, with the knowledge that the sinusoidal form can be recovered at the end of any calculation. In fact, so secure are we in the one-to-one correspondence between the two expressions that often the sinusoidal form is not recovered. It is, however, salutary to remember that an equation such as

$$v = V\,e^{j(\omega t + \phi)}$$

cannot be plotted on a graph of v against t.

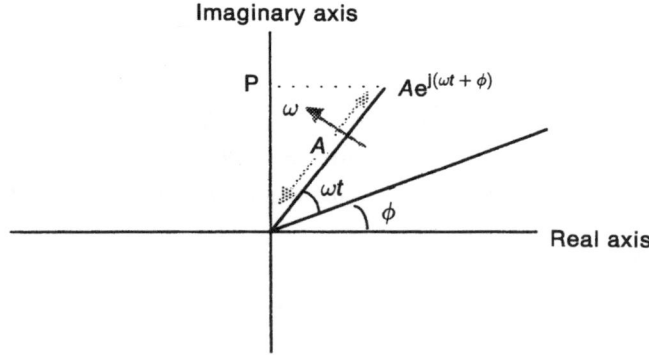

Fig. A1.2   Representation on an Argand diagram of the rotating line joining the origin to $A\,e^{j(\omega t + \phi)}$

Appendix 2
Inductance and capacitance of transmission line

A2.1 TWIN LINE

Consider the line consisting of the pair of wires shown in Fig. A2.1; each wire has radius a, and the separation between the centre of one wire and the surface of the other is b. The current in the wires is I, into the page for the wire marked A and out of the page for the one marked B. The voltage between the wires is V.

Inductance

The magnetic field due to the current in A encircles A and has a strength at radius r given by

$$H_r = \frac{I}{2\pi r} \ \text{A/m}$$

Hence the magnetic flux density at radius r

$$B_r = \frac{\mu_0 I}{2\pi r} \ \text{Wb/m}^2$$

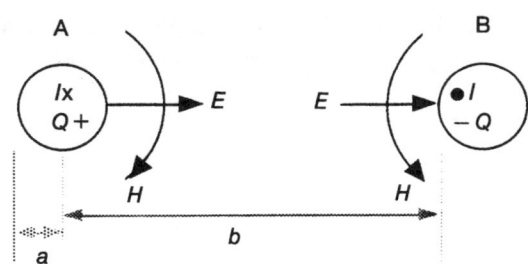

Fig. A2.1 Cross-section of a twin-wire line with fields. Current into A and out of B

The total flux linking the space between the wires, per metre length of line, due to the current in A is

$$\Phi_A = \int_a^b \frac{\mu_0 I}{2\pi r}\, dr \text{ Wb/m}$$

which yields

$$\Phi_A = \frac{\mu_0 I}{2\pi} \ln\frac{b}{a} \text{ Wb/m}$$

(We have ignored any flux within wire A, which is reasonable, since a.c. current will flow near the surface due to the skin effect.)

The flux due to B will have the same value and will add to that of A between the wires, since the currents are in opposite directions, so the total flux linkage per metre

$$\Phi = \frac{\mu_0 I}{\pi} \ln\frac{b}{a} \text{ Wb/m} \tag{A2.1}$$

The inductance is the flux linkage per unit current, so

$$L = \frac{\mu_0}{\pi} \ln\frac{b}{a} \text{ H/m} \tag{A2.2}$$

Capacitance

Let the two lines carry opposite charges, $+Q$ coulombs on A and $-Q$ coulombs on B, per metre length.

The electric flux density at radius r outward from line A is

$$D_A = \frac{Q}{2\pi r} \text{ C/m}^2$$

so the electric field strength, which is D/ε, is

$$E_A = \frac{Q}{2\pi r \varepsilon} \text{ V/m}$$

The voltage between the wires due to the charge on A is then

$$V_A \approx \int_a^b E_A\, dr = \frac{Q}{2\pi\varepsilon} \ln\frac{b}{a} \text{ V}$$

The voltage due to the charge on B has the same value, and, since the electric field direction is inwards towards B, acts in the same sense as V_A, so the total voltage between the lines is

$$V = \frac{Q}{\pi\varepsilon} \ln\frac{b}{a} \text{ V} \tag{A2.3}$$

The capacitance is Q/V, and since Q is charge per unit length,

$$C = \frac{\pi\varepsilon}{\ln(b/a)} \text{ F/m} \tag{A2.4}$$

Phase velocity

In Chapter 2 the phase velocity on an ideal line (or a line at high frequency) is shown to be $1/\sqrt{LC}$, which, from Equations (A2.2) and (A2.4) can be seen to have the value

$$v_p = 1/\sqrt{\mu_0 \varepsilon}$$

Characteristic impedance

Again from Chapter 2, the characteristic impedance is $\sqrt{L/C}$, which, from Equations (A2.2) and (A2.4) gives for the characteristic impedance in terms of the line dimensions

$$Z_0 = \frac{\ln(b/a)}{\pi} \sqrt{\frac{\mu_0}{\varepsilon}}$$

A2.2   COAXIAL LINE

Figure A2.2 shows a cross-section of a coaxial line with an inner conductor of radius a and an outer conductor with an inside radius b. The currents in the conductors have value I, in opposite directions, and the voltage between the conductors is V.

Inductance

All the magnetic field between the conductors is due to the current in the inner conductor. Outside the outer conductor the net field is zero because the field due to the current in the outer conductor is equal and opposite to that due to the current in the inner conductor.

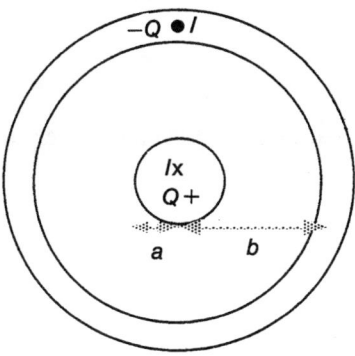

Fig. A2.2   Cross-section of a coaxial line. Current into the inner conductor and out of the outer conductor

The field strength at radius r between the conductors is

$$H_r = \frac{I}{2\pi r} \text{ A/m}$$

Hence the magnetic flux density at radius r

$$B_r = \frac{\mu_0 I}{2\pi r} \text{ Wb/m}^2$$

The total magnetic flux between the conductors is

$$\Phi = \int_a^b \frac{\mu_0 I}{2\pi r} \, dr \text{ Wb/m}$$

which yields

$$\Phi = \frac{\mu_0 I}{2\pi} \ln \frac{b}{a} \text{ Wb/m}$$

The inductance is the flux linkage per unit current, so

$$L = \frac{\mu_0}{2\pi} \ln \frac{b}{a} \text{ H/m} \tag{A2.5}$$

Capacitance

Let the two lines carry opposite charges, $+Q$ coulombs and $-Q$ coulombs per metre length. Again, all the electric field will be in the space between inner and outer conductors.

The electric flux density at radius r is

$$D = \frac{Q}{2\pi r} \text{ C/m}^2$$

so the electric field strength, which is D/ε, is

$$E = \frac{Q}{2\pi r \varepsilon} \text{ V/m}$$

The voltage between the conductors is then

$$V \approx \int_a^b E \, dr = \frac{Q}{2\pi \varepsilon} \ln \frac{b}{a} \text{ V}$$

The capacitance is Q/V, and since Q is charge per unit length,

$$C = \frac{2\pi \varepsilon}{\ln(b/a)} \text{ F/m} \tag{A2.6}$$

Phase velocity

Again using $1/\sqrt{LC}$, from Equations (A2.5) and (A2.6)

$$v_p = 1/\sqrt{\mu_0 \varepsilon}$$

Characteristic impedance

From Equations (A2.5) and (A2.6) $\sqrt{L/C}$ yields

$$Z_0 = \frac{\ln(b/a)}{2\pi} \sqrt{\frac{\mu_0}{\varepsilon}}$$

A2.3   MICROSTRIP

Figure A2.3 shows a cross-section of a microstrip (a repeat of Fig. 4.2). Mathematical analysis of the fields produced by this configuration is much more difficult than for twin-wire line or coax; numerical techniques exist which are best applied by a computer. From the previous two analyses it is reasonable to deduce that the phase velocity in microstrip will be $1/\sqrt{\mu_0 \varepsilon}$.

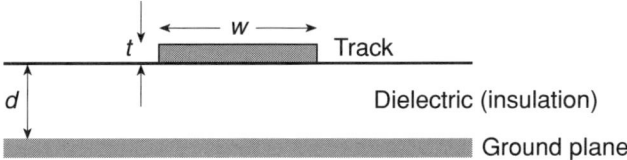

Fig. A2.3   Microstrip

A clue to the nature of Z_0 for microstrip can be found by considering a cylindrical wire above a ground plane as shown in Fig. A2.4. Using the method of electrical images, the wire will have an image in the ground plane so that the same method of analysis can be used as in Section A2.1, but the magnetic flux linking the circuit only exists between the wire and the ground plane and so is only half the value given in Equation (A2.1). Also, $d \approx b/2$ and $w = 2a$, so we can write

$$L \approx \frac{\mu_0}{2\pi} \ln \frac{4d}{w} \text{ H/m} \tag{A2.7}$$

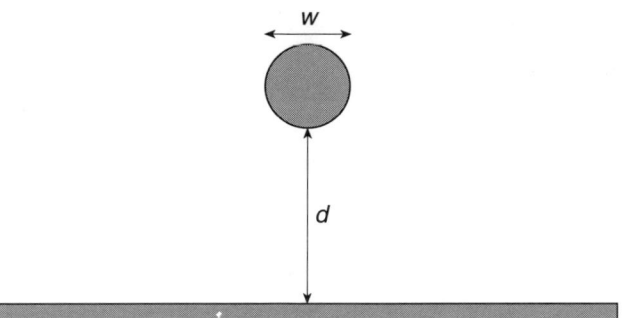

Fig. A2.4   Cylindrical conductor above a ground plane

The voltage between the wire and the ground plane will be half the value given in Equation (A2.3) giving

$$C \approx \frac{2\pi\varepsilon}{\ln(4d/w)} \text{ F/m} \tag{A2.8}$$

Again it will be seen that $v_{\mathrm{p}} = 1/\sqrt{\mu_0\varepsilon}$ and here

$$Z_0 \approx \frac{\ln(4d/w)}{2\pi} \sqrt{\frac{\mu_0}{\varepsilon}}$$

Appendix 3
The wave-propagating properties of media

The aim here is to establish properties that a medium must possess in order to propagate waves of the sort described in the main text. Invoking Fourier's theorem, we take the waveform to consist of an arbitrary set of sine waves.

Consider some property of a medium of which the instantaneous value can be represented by a. If a loss-free wave of the variation of this property will propagate in one dimension in the medium, then its value at distance x in the direction of propagation from some reference point and at time t after some reference time can be represented as

$$a = A_1 \sin(\omega_1 t - \beta_1 x) + A_2 \sin(\omega_2 t - \beta_2 x) + A_3 \sin(\omega_3 t - \beta_3 x) + \cdots$$

We now differentiate this expression twice with respect to x and twice with respect to t.

With respect to x

$$\frac{\partial a}{\partial x} = -\beta_1 A_1 \cos(\omega_1 t - \beta_1 x) - \beta_2 A_2 \cos(\omega_2 t - \beta_2 x)$$
$$- \beta_3 A_3 \cos(\omega_3 t - \beta_3 x) - \cdots$$

$$\frac{\partial^2 a}{\partial x^2} = -\beta_1^2 A_1 \sin(\omega_1 t - \beta_1 x) - \beta_2^2 A_2 \sin(\omega_2 t - \beta_2 x)$$
$$- \beta_3^2 A_3 \sin(\omega_1 t - \beta_3 x) - \cdots$$

(A3.1)

With respect to t

$$\frac{\partial a}{\partial t} = \omega_1 A_1 \cos(\omega_1 t - \beta_1 x) + \omega_2 A_2 \cos(\omega_2 t - \beta_2 x)$$
$$+ \omega_3 A_3 \cos(\omega_3 t - \beta_3 x) + \cdots$$

$$\frac{\partial^2 a}{\partial t^2} = -\omega_1^2 A_1 \sin(\omega_1 t - \beta_1 x) - \omega_2^2 A_2 \sin(\omega_2 t - \beta_2 x)$$
$$- \omega_3^2 A_3 \sin(\omega_1 t - \beta_3 x) - \cdots$$

(A3.2)

Equation (A3.1) can be rewritten

$$\frac{\partial^2 a}{\partial x^2} = -\frac{\beta_1^2}{\omega_1^2}[\omega_1^2 A_1 \sin(\omega_1 t - \beta_1 x)] - \frac{\beta_2^2}{\omega_2^2}[\omega_2^2 A_2 \sin(\omega_2 t - \beta_2 x)]$$

$$-\frac{\beta_3^2}{\omega_3^2}[\omega_3^2 A_3 \sin(\omega_3 t - \beta_3 x)] - \cdots \tag{A3.3}$$

Now, provided β is proportional to ω, we can write

$$\frac{\beta_1^2}{\omega_1^2} = \frac{\beta_2^2}{\omega_2^2} = \frac{\beta_3^2}{\omega_3^2} = \cdots = \frac{1}{v_p^2}$$

where v_p is the phase velocity for any sinusoid, and rewrite Equation (A3.3)

$$\frac{\partial^2 a}{\partial x^2} = \frac{1}{v_p^2}[-\omega_1^2 A_1 \sin(\omega_1 t - \beta_1 x) - \omega_2^2 A_2 \sin(\omega_2 t - \beta_2 x)$$

$$- \omega_3^2 A_3 \sin(\omega_1 t - \beta_3 x) - \cdots] \tag{A3.4}$$

From Equations (A3.2) and (A3.4)

$$\frac{\partial^2 a}{\partial x^2} = \frac{1}{v_p^2}\frac{\partial^2 a}{\partial t^2} \tag{A3.5}$$

What has been shown (in a back-to-front way) is that the sum of a series of sine-waves of the form $A_n \sin(\omega_n t - \beta_n x)$ is a solution of the differential Equation (A3.5) provided the ratio β_n/ω_n is constant, and so if a property of a medium can be shown to obey an equation of the form of (A3.5) then it will propagate a one-dimensional wave of any waveform without loss and without dispersion.

Provided a medium is isotropic – that is, has the same properties in every direction – analysis of its wave-propagating properties in one dimension is sufficient to find the phase velocity in terms of the properties of the medium.

For a wave spreading out in three dimensions in a medium, it can be shown that the appropriate differential equation is

$$\frac{\partial^2 a}{\partial x^2} + \frac{\partial^2 a}{\partial y^2} + \frac{\partial^2 a}{\partial z^2} = \frac{1}{v_p^2}\frac{\partial^2 a}{\partial t^2}$$

A3.1 ANALYSIS OF A TRANSMISSION LINE

Look at Fig. A3.1. This is similar to Fig. 2.5, but now v and i are instantaneous current and voltage. We shall undertake an analysis from first principles rather than using phasors as in Chapter 2.

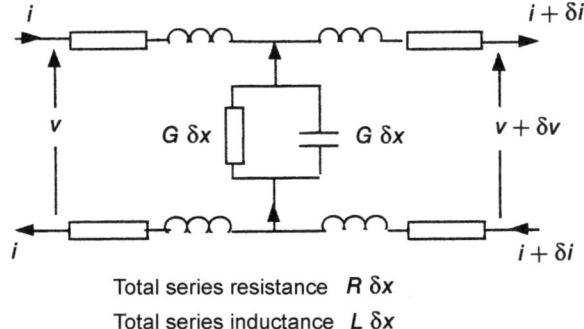

Total series resistance $R\,\delta x$

Total series inductance $L\,\delta x$

Fig. A3.1   Electrical properties of a length of line δx

Applying Kirchhoff's voltage law around the loop of the wires, and ignoring the small current variation

$$\delta v = -R\,\delta x i - L\,\delta x \frac{\partial i}{\partial t}$$

Dividing by δx and allowing $\delta x \to 0$

$$\frac{\partial v}{\partial x} = -Ri - L\frac{\partial i}{\partial t} \tag{A3.6}$$

Similarly, considering the current between the wires and ignoring the small voltage variation

$$\delta i = -G\,\delta x v - C\,\delta x \frac{\partial v}{\partial t}$$

giving

$$\frac{\partial i}{\partial x} = -Gv - C\frac{\partial v}{\partial t} \tag{A3.7}$$

Differentiating Equation (A3.6) with respect to x

$$\frac{\partial^2 v}{\partial x^2} = -R\frac{\partial i}{\partial x} - L\frac{\partial^2 i}{\partial t\,\partial x} \tag{A3.8}$$

and Equation (A3.7) with respect to t

$$\frac{\partial^2 i}{\partial x\,\partial t} = -G\frac{\partial v}{\partial t} - C\frac{\partial^2 v}{\partial t^2} \tag{A3.9}$$

Combining Equations (A3.7), (A3.8) and (A3.9)

$$\frac{\partial^2 v}{\partial x^2} = RGv + RC\frac{\partial v}{\partial t} + LG\frac{\partial v}{\partial t} + LC\frac{\partial^2 v}{\partial t^2}$$

and rearranging

$$\frac{\partial^2 v}{\partial x^2} = LC\frac{\partial^2 v}{\partial t^2} + (RC + LG)\frac{\partial v}{\partial t} + RGv \tag{A3.10}$$

An exactly similar equation can be derived for the current in the line, i.e.

$$\frac{\partial^2 i}{\partial x^2} = LC\frac{\partial^2 i}{\partial t^2} + (RC + LG)\frac{\partial i}{\partial t} + RGi \qquad (A3.11)$$

If R and G are both zero, then we have a loss-free, dispersion free, transmission line, characterized by the differential equations

$$\frac{\partial^2 v}{\partial x^2} = LC\frac{\partial^2 v}{\partial t^2} \qquad \text{and} \qquad \frac{\partial^2 i}{\partial x^2} = LC\frac{\partial^2 i}{\partial t^2}$$

indicating that a wave with no attenuation, that can be described in terms of either the voltage between the conductors of the line or the current in the conductors of the line, will propagate with a phase velocity $1/\sqrt{LC}$.

Where equations with extra terms similar to those in Equations (A3.10) and (A3.11) occur, the indication is that an attenuated wave will propagate, and, except under special circumstances, it will be dispersive.

A3.2   ANALYSIS OF A LOSS-FREE DIELECTRIC MEDIUM

The electric and magnetic properties of a loss-free dielectric medium are characterized by Maxwell's equations, which, assuming there is no stored charge in the medium and no conduction current, can be written

$$\text{div } \boldsymbol{D} = 0 \qquad (A3.12)$$

$$\text{div } \boldsymbol{B} = 0 \qquad (A3.13)$$

$$\text{curl } \boldsymbol{E} = -\frac{\partial \boldsymbol{B}}{\partial t} \qquad (A3.14)$$

$$\text{curl } \boldsymbol{H} = \frac{\partial \boldsymbol{D}}{\partial t} \qquad (A3.15)$$

Also, assume that for the medium

$$\boldsymbol{D} = \varepsilon \boldsymbol{E} \quad \text{where} \quad \varepsilon = \varepsilon_\mathrm{r}\varepsilon_0 \qquad (A3.16)$$

and

$$\boldsymbol{B} = \mu_0 \boldsymbol{H} \qquad (A3.17)$$

For details of Maxwell's equations, see *Electromagnetism*, 2nd Edition by I.S. Grant and W.R. Phillips published by John Wiley and Sons.

In rectangular cartesian coordinates as shown in Fig. A3.2 (a repeat of Fig. 3.1) Equation (A3.12) becomes

$$\frac{\partial D_x}{\partial x} + \frac{\partial D_y}{\partial y} + \frac{\partial D_z}{\partial z} = 0 \qquad (A3.18)$$

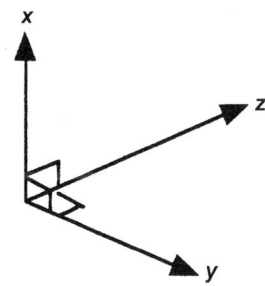

Fig. A3.2   Three-dimensional coordinates

while Equation (A3.13) becomes

$$\frac{\partial B_x}{\partial x} + \frac{\partial B_y}{\partial y} + \frac{\partial B_z}{\partial z} = 0 \tag{A3.19}$$

Equation (A3.14) becomes three equations

$$\frac{\partial E_z}{\partial y} - \frac{\partial E_y}{\partial z} = -\frac{\partial B_x}{\partial t}$$

$$\frac{\partial E_x}{\partial z} - \frac{\partial E_z}{\partial x} = -\frac{\partial B_y}{\partial t} \tag{A3.20}$$

$$\frac{\partial E_y}{\partial x} - \frac{\partial E_x}{\partial y} = -\frac{\partial B_z}{\partial t}$$

and Equation (A3.15) also becomes three equations

$$\frac{\partial H_z}{\partial y} - \frac{\partial H_y}{\partial z} = \frac{\partial D_x}{\partial t}$$

$$\frac{\partial H_x}{\partial z} - \frac{\partial H_z}{\partial x} = \frac{\partial D_y}{\partial t} \tag{A3.21}$$

$$\frac{\partial H_y}{\partial x} - \frac{\partial H_x}{\partial y} = \frac{\partial D_z}{\partial t}$$

Assuming that any electric field in the medium is in one direction only, which we take as the x direction, then Equation (A3.18) reduces to

$$\frac{\partial D_x}{\partial x} = 0$$

so that, using Equation (A3.16)

$$\frac{\partial E_x}{\partial x} = 0$$

indicating that such an electric field must have a constant value at all points in the x direction (consistent with the idea that electric flux can only terminate on charge). Assuming also that the electric field strength is uniform over the x/y

plane, so that $\partial D_x/\partial y$ is also zero, in which case Equation (A3.20) reduces to

$$\frac{\partial B_x}{\partial t} = 0 \qquad \frac{\partial B_z}{\partial t} = 0 \qquad \text{and} \qquad -\frac{\partial B_y}{\partial t} = \frac{\partial E_x}{\partial z} \tag{A3.22}$$

so the only time-varying magnetic field is B_y, and using Equation (A3.17) we can write

$$\frac{\partial E_x}{\partial z} = -\mu_0 \frac{\partial H_y}{\partial t} \tag{A3.23}$$

One of the components of Equation (A3.21) is

$$\frac{\partial H_z}{\partial y} - \frac{\partial H_y}{\partial z} = \frac{\partial D_x}{\partial t} \tag{A3.24}$$

We have seen in (A3.22) that there is no time-varying component of B_z – and hence of H_z – so, ignoring static fields, Equations (A3.24) and (A3.16) together yield

$$\frac{\partial H_y}{\partial z} = -\varepsilon \frac{\partial E_x}{\partial t} \tag{A3.25}$$

Differentiating Equation (A3.23) with respect to z gives

$$\frac{\partial^2 E_x}{\partial z^2} = -\mu_0 \frac{\partial^2 H_y}{\partial t\, \partial z} \tag{A3.26}$$

and differentiating Equation (A3.25) with respect to t

$$\frac{\partial^2 H_y}{\partial z\, \partial t} = -\varepsilon \frac{\partial^2 E_x}{\partial t^2} \tag{A3.27}$$

and thus, from Equations (A3.26) and (A3.27)

$$\frac{\partial^2 E_x}{\partial z^2} = \mu_0\varepsilon \frac{\partial^2 E_x}{\partial t^2} \tag{A3.28}$$

and in an exactly similar manner

$$\frac{\partial^2 H_y}{\partial z^2} = \mu_0\varepsilon \frac{\partial^2 H_y}{\partial t^2} \tag{A3.29}$$

Equations (A3.28) and (A3.29) indicate that a plane wave, of the sort described in Chapter 3, will propagate in a loss-free dielectric medium with a phase velocity

$$v_p = 1/\sqrt{\mu_0\varepsilon} \tag{A3.30}$$

Let us now consider a sinusoidal electric field which we shall designate

$$E_x = E_m \sin(\omega t - \beta z) \tag{A3.31}$$

Differentiating Equation (A3.31) with respect to z gives

$$\frac{\partial E_x}{\partial z} = -\beta E_m \cos(\omega t - \beta z) \tag{A3.32}$$

Thus, from Equations (A3.23) and (A3.32)

$$\frac{\partial H_y}{\partial t} = \frac{\beta}{\mu_0} E_m \cos(\omega t - \beta z)$$

and integrating this with respect to t gives

$$H_y = \frac{\beta}{w\mu_0} E_m \sin(\omega t - \beta z) \qquad\qquad (A3.33)$$

From Equations (A3.31) and (A3.33)

$$\frac{E_x}{H_y} = \frac{\omega\mu_0}{\beta} \qquad\qquad (A3.34)$$

Equation (A3.30) indicates that $\omega/\beta = 1/\sqrt{\mu_0\varepsilon}$, so (A3.34) becomes

$$\frac{E_x}{H_y} = \sqrt{\frac{\mu_0}{\varepsilon}}$$

This ratio is known as the *wave impedance*.

Appendix 4
Algebraic analysis of the fundamental mode in rectangular waveguide

Figure A4.1 shows a rectangular waveguide and its orientation with respect to x, y, z axes. In order to develop equations relating to the fundamental mode, we shall have to make certain assumptions regarding the distribution of electric field obtained from the graphical analysis given in Chapter 4; thus we shall assume that $E_y = 0$, that $E_z = 0$ and that the value of E_x does not vary in the x direction, so that $\partial^2 E_x / \partial x^2 = 0$.

Hence, for the medium contained in the guide (which we shall assume to be air, with a relative permittivity negligibly different from unity), we can write

$$\frac{\partial^2 E_x}{\partial y^2} + \frac{\partial^2 E_x}{\partial z^2} = \frac{1}{c^2}\frac{\partial^2 E_x}{\partial t^2} \tag{A4.1}$$

(*see* Appendix 3).

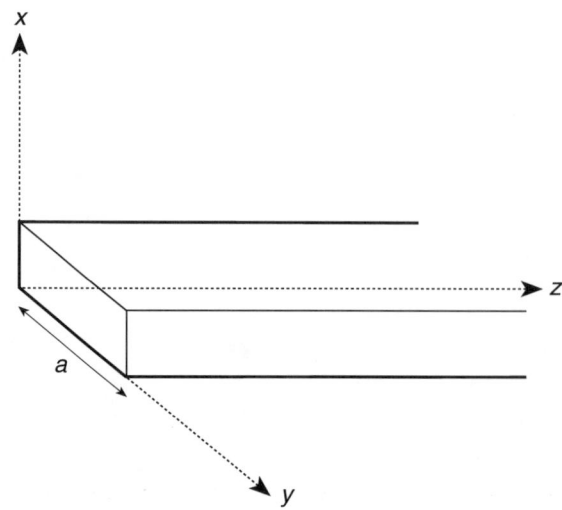

Fig. A4.1   A waveguide with coordinate axes

Since propagation is in the z direction, we can assume that E_x shows no phase variation in the y direction. Also, the boundary conditions indicate that $E_x = 0$ when $y = 0$ and when $y = a$.

Again, the geometrical analysis suggests that the distribution of E_x amplitude between $y = 0$ and $y = a$ might be a half-cycle of a sinusoid, so we shall try for a solution of Equation (A4.1) of the form

$$E_x = A \sin\left(\frac{\pi y}{a}\right) e^{(j\omega t - \gamma_g z)} \tag{A4.2}$$

where γ_g is the propagation constant in the guide. (*See* Appendix 1 for the exponential representation.)

Differentiating Equation (A4.2) yields:

$$\frac{\partial^2 E_x}{\partial y^2} = -\left(\frac{\pi}{a}\right)^2 E_x \tag{A4.3}$$

$$\frac{\partial^2 E_x}{\partial z^2} = \gamma_g^2 E_x \tag{A4.4}$$

$$\frac{\partial^2 E_x}{\partial t^2} = (j\omega)^2 E_x = -\omega^2 E_x \tag{A4.5}$$

Substituting (A4.3), (A4.4) and (A4.5) in (A4.1) and cancelling E_x gives

$$\gamma_g^2 = \left(\frac{\pi}{a}\right)^2 - \left(\frac{\omega}{c}\right)^2$$

which, taking $2a$ as λ_c, can be written

$$\gamma_g^2 = 4\pi^2 \left(\frac{1}{\lambda_c^2} - \frac{1}{\lambda^2}\right) \tag{A4.6}$$

From Equation (A4.6) it can be seen that γ_g is purely real or purely imaginary according as $1/\lambda^2$ is less than or greater than $1/\lambda_c^2$, i.e. whether f is less than or greater than f_c.

Taking first the case where $f > f_c$.

$$\gamma_g = 2\pi \sqrt{-\left(\frac{1}{\lambda^2} - \frac{1}{\lambda_c^2}\right)}$$

$$\gamma_g = j2\pi \sqrt{\frac{1}{\lambda^2} - \frac{1}{\lambda_c^2}}$$

Now we can write

$$\gamma_g = \alpha_g + j\beta_g$$

So in this case $\alpha_g = 0$.

(But note: this does not take account of resistive losses in the waveguide walls, because we have not included these in the boundary conditions when setting up the equations.)

$$\beta_g = 2\pi\sqrt{\frac{1}{\lambda^2} - \frac{1}{\lambda_c^2}}$$

$$\left(\frac{\beta_g}{2\pi}\right)^2 = \frac{1}{\lambda^2} - \frac{1}{\lambda_c^2}$$

but

$$\frac{2\pi}{\beta_g} = \lambda_g$$

so, when $f > f_c$

$$\frac{1}{\lambda_g^2} = \frac{1}{\lambda^2} - \frac{1}{\lambda_c^2}$$

which is Equation (4.4) from Chapter 4.

When $f < f_c$

$$\beta_g = 0$$

$$\alpha_g = 2\pi\sqrt{\frac{1}{\lambda_c^2} - \frac{1}{\lambda^2}}$$

$$\frac{1}{\lambda_c^2} = \frac{f_c^2}{c^2} \quad \text{and} \quad \frac{1}{\lambda^2} = \frac{f^2}{c^2}$$

so

$$\alpha_g = \frac{2\pi}{c}\sqrt{f_c^2 - f^2}$$

There is no phase change with distance along the guide, but the fields decay exponentially with distance along the guide. This is known as *evanescent propagation*.

The attenuation in a cut-off waveguide is not attenuation in the sense of wave energy turned into heat, it simply represents an exponential fall off of field amplitudes: such field energy as is not able to penetrate to some energy absorbing termination is reflected back to the source. A short section of cut-off waveguide between two propagating guides will pass through a proportion of the incident energy depending on the length of the cut-off section; such an arrangement is sometimes used as an attenuator.

Evanescent propagation into the cladding of an optical fibre in the direction at right angles to the fibre axis is a similar phenomenon.

Wave impedance

A relevant component of Maxwell's equations, (A3.20) in Appendix 3, is

$$\frac{\partial E_x}{\partial z} - \frac{\partial E_z}{\partial x} = -\frac{\partial B_y}{\partial t} \tag{A4.7}$$

Differentiating Equation (A4.2) and using $B_y = \mu_0 H_y$, Equation (A4.7) becomes

$$\mu_0 \frac{\partial H_y}{\partial t} = \gamma_g A \sin\left(\frac{\pi y}{a}\right) e^{(j\omega t - \gamma_g z)}$$

Integrating this with respect to time

$$H_y = \frac{\gamma_g}{j\omega\mu_0} A \sin\left(\frac{\pi y}{a}\right) e^{(j\omega t - \gamma_g z)} \tag{A4.8}$$

Dividing (A4.2) by (A4.8) yields

$$\frac{E_x}{H_y} = \frac{j\omega\mu_0}{\gamma_g} \tag{A4.9}$$

In the frequency range of wave propagation

$$\gamma_g = j\beta_g = \frac{j2\pi}{\lambda_g}$$

and, using also $\omega = 2\pi c/\lambda$, Equation (A4.9) becomes

$$\frac{E_x}{H_y} = \mu_0 c \frac{\lambda_g}{\lambda} \tag{A4.10}$$

$$\mu_0 c = \frac{\mu_0}{\sqrt{\mu_0 \varepsilon_0}} = \sqrt{\frac{\mu_0}{\varepsilon_0}}$$

which is the wave impedance of free space (value 120π) and which we shall call Z_f.

Equation (A4.10) shows that the ratio of the transverse components of electric and magnetic fields in the fundamental mode in a waveguide is everywhere the same: it is known as the *guide wave impedance* Z_W.

$$Z_W = Z_f \frac{\lambda_g}{\lambda}$$

Appendix 5
The polar diagram of a radiating aperture

Consider first a line of n radiating point sources, separated by a distance d, all of equal power and in phase – see Fig. A5.1. At a distant point, in a direction making an angle θ to the normal to the line, each can be considered to produce the same amplitude of electric field, but there is a small path-length difference between the radiation from adjacent sources, marked x in Fig. A5.1. (The distance to the receiving point is large enough compared to the length of the array of sources that the lines joining each source to it are effectively parallel as shown.)

$x = d \sin \theta$

The phase difference associated with this path-length difference we shall call ϕ, where

$$\phi = \frac{2\pi x}{\lambda} = \frac{2\pi d \sin \theta}{\lambda} \tag{A5.1}$$

The electric field at the distant point due to the rth source from the right in the figure can be written

$E_r = E_1 \, e^{-jr\phi}$

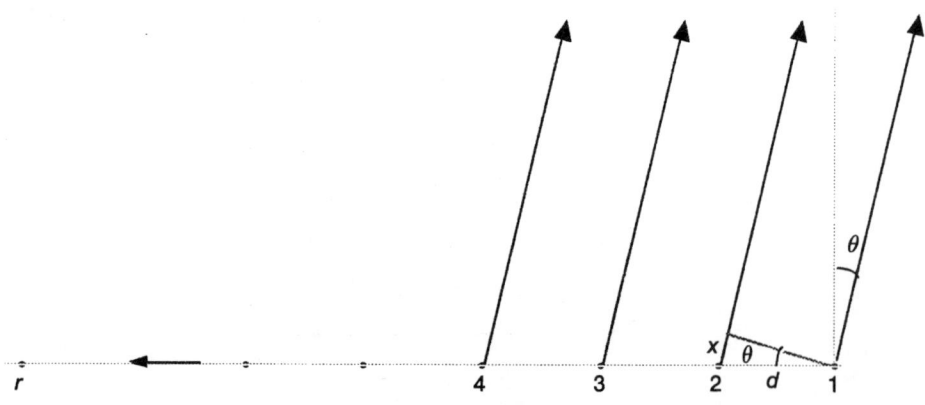

Fig. A5.1   A row of point sources radiating in phase

(*See* Appendix 1 for an explanation of this exponential representation of phase delay.)

So, the total electric field due to all the sources in the direction θ is

$$E_\theta = E_1 + E_1\,e^{-j\phi} + E_1\,e^{-j2\phi} + \cdots + E_1\,e^{-j(n-1)\phi}$$

This is a geometric progression with a common ratio $e^{-j\phi}$ which can be summed to give

$$E_\theta = E_1\frac{(1 - e^{-jn\phi})}{1 - e^{-j\phi}} = E_1\frac{e^{-jn\phi/2}}{e^{-j\phi/2}}\,\frac{e^{jn\phi/2} - e^{-jn\phi/2}}{e^{j\phi/2} - e^{-j\phi/2}} \tag{A5.2}$$

$$e^{jn\phi/2} - e^{-jn\phi/2} = 2\sin\frac{n\phi}{2} \tag{A5.3}$$

and

$$e^{j\phi/2} - e^{-j\phi/2} = 2\sin\frac{\phi}{2} \tag{A5.4}$$

Also

$$\frac{e^{-jn\phi/2}}{e^{-j\phi/2}} = e^{-j(n-1)\phi/2}$$

which is the average phase of the total field.

So, from (A5.1), (A5.2), (A5.3) and (A.5.4) we can write for the magnitude of the total electric field at the distant point at angle θ

$$|E_\theta| = |E_1|\frac{\sin\left(\dfrac{n\pi d\sin\theta}{\lambda}\right)}{\sin\left(\dfrac{\pi d\sin\theta}{\lambda}\right)} \tag{A5.5}$$

Equation (A5.5) is the equation for the polar diagram of a broadside array of dipoles, for example.

Going on now to consider an aperture, imagine that the aperture is filled with a matrix of point sources, all separated by a distance d, as shown in Fig. A5.2. If we consider the polar diagram in a plane perpendicular to the aperture and parallel to

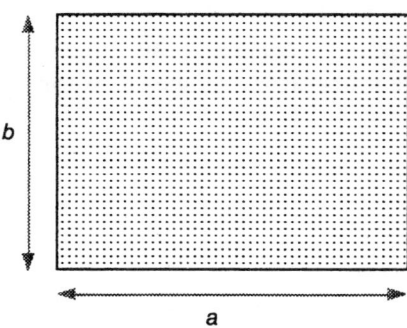

Fig. A5.2   An array of point sources

the side a, then Equation (A5.5) will still hold because the radiation from a column of sources parallel to side b are all in phase ($|E_1|$ will now represent the field magnitude due to all the sources in a column rather than one source).

But now we can write

$$nd = a$$

so Equation (A5.5) can be rewritten

$$|E_\theta| = |E_1| \frac{\sin\left(\dfrac{\pi a \sin\theta}{\lambda}\right)}{\sin\left(\dfrac{\pi a \sin\theta}{n\lambda}\right)} \tag{A5.6}$$

If we let d tend to zero and n tend to infinity, giving an aperture radiating from all points in phase, then, since the sine of a small angle tends to the value of the angle (in radians) as the angle tends to zero, Equation (A5.6) becomes

$$|E_\theta| = n|E_1| \frac{\sin\left(\dfrac{\pi a \sin\theta}{\lambda}\right)}{\dfrac{\pi a \sin\theta}{\lambda}} \tag{A5.7}$$

$n|E_1|$ is the electric field magnitude due to the whole aperture, which will of course be finite, and which we shall now write $|E_m|$. a/λ is the aperture in wavelengths, so it is convenient to link these two symbols together. Finally then, for the polar diagram of the aperture in the plane of a,

$$|E_\theta| = |E_m| \frac{\sin\{\pi(a/\lambda)\sin\theta\}}{\pi(a/\lambda)\sin\theta} \tag{A5.8}$$

The polar diagram in the plane of b will be exactly the same, with b substituted for a.

The expression

$$\frac{\sin\{\pi(a/\lambda)\sin\theta\}}{\pi(a/\lambda)\sin\theta}$$

is unity when $\theta = 0$ and is less than unity for all other values of θ, so $|E_m|$ is the maximum amplitude. With increasing θ, the value of the expression falls until it becomes zero when the numerator becomes zero for

$$\pi(a/\lambda)\sin\theta = \pi$$

giving

$$\sin\theta = \lambda/a \tag{A5.9}$$

The magnitude of the expression then rises and falls again to the next zero when

$$\pi(a/\lambda)\sin\theta = 2\pi$$

giving

$$\sin \theta = 2\lambda/a$$

and so on as long as $n\lambda/a$ is less than unity. This gives a pattern of a main lobe with side lobes of rapidly decreasing peak value, as shown in Fig. 6.9.

If λ/a is small (the aperture is many wavelengths across) then, from Equation (A5.9), approximately the first zero on either side of the main lobe comes at an angle λ/a radians, so the angle between zeros is $2\lambda/a$: the beamwidth is defined as the angle between half-power directions, and this is approximately λ/a radians as indicated in Fig. 6.9.

For a circular aperture the analysis is not so straightforward. The result is a polar diagram of exactly the same form as for a square aperture, but the first zeros occur at an angle that is about 25 per cent greater. When dealing with a paraboloidal antenna one has to make an approximation for the effective aperture anyway, so the rule 'beamwidth is the reciprocal of effective diameter in wavelengths' is still applied.

Solutions to calculations

2.1 Z_0 as quoted by the manufacturer is equal to $\sqrt{L/C}$ so

$$L = CZ_0^2 = 65 \times 10^{-12} \times 78^2 = \mathbf{395\,nH/m}$$

2.2 At 1 MHz the change in amplitude over a distance of 100 m, corresponding to $-2\,\mathrm{dB}$ (minus because it is attenuation) equals

$$10^{-2/20} = 0.794$$

From the definition of attenuation coefficient this amplitude ratio equals $e^{-\alpha x}$, so

$$e^{-100\alpha} = 0.794$$

Taking logs to base e

$$-100\alpha = \ln 0.794 = -0.231$$

so,

$$\alpha = \mathbf{2.31 \times 10^{-3}\,N/m}$$

2.3 $\alpha = \frac{1}{2}(R/R_0 + G/G_0)$. If G is negligible

$$R = 2\alpha R_0 = 2 \times 2.31 \times 10^{-3} \times 78 = 0.36\,\Omega/\mathrm{m}$$

This is the highest possible value of R.

$$\omega L = 2\pi \times 10^6 \times 395 \times 10^{-9} = 2.48$$

which is about 7 times $R_{\max}$. If R is negligible

$$G = 2\alpha G_0 = 2 \times 2.31 \times 10^{-3} \times 1/78 = 5.9 \times 10^{-5}\,\mathrm{S/m}$$

$\omega C = 2\pi \times 10^6 \times 65 \times 10^{-12} = 4 \times 10^{-4}$ which again is about 7 times $G_{\max}$.
   Since the attenuation will be shared between series resistance and shunt conductance we are justified in assuming that ωL and ωC are an order of magnitude greater than R and G respectively so the line can be considered non-dispersive at 1 MHz.

2.4 The data show that the line is becoming more lossy as frequency increases, but the loss per 100 m only increases from 2 to 16.4, that is 8.2 times, in decibels when the frequency increases 50 times. Since nepers are proportional to decibels, α also increases 8.2 times (notice that this is not much more than the square root of the frequency increase). As R and G depend on α, this indicates that the line will not be dispersive above 1 MHz. (Since the attenuation can be expected to be less at lower frequencies, the line is not necessarily dispersive at frequencies somewhat below 1 MHz either.)

2.5 The phase velocity on the line at all three frequencies quoted will be

$$1\sqrt{LC} = 1\sqrt{(395 \times 10^{-9}) \times (65 \times 10^{-12})} = \mathbf{1.97 \times 10^8 \, m/s}$$

(The velocity is less than 3×10^8 m/s because the wires are insulated with polythene.)

2.6 The wavelength on the line at 10 MHz is

$$v_{\mathrm{p}}/f = 1.97 \times 10^8/10^7 = \mathbf{19.7 \, m}$$

2.7 $\beta = \omega/v_{\mathrm{p}} = 2\pi \times 50 \times 10^6/1.97 \times 10^8 = 1.5947 \, \mathrm{rad/m}$. The phase delay over 10 m is therefore

$$15.947 \, \mathrm{rad} = 15.947 \times 180/\pi = 913.7°$$

Two full cycles is $2 \times 360° = 720°$, so the lag of the sinusoidal variation at the end of 10 m compared to the beginning is $913.7° - 720° = 193.7°$.

The phase difference, which is usually expressed as the smaller angle between the phases is

$$360° - 193.7° = \mathbf{166.3°}$$

2.8 The impedance of R in parallel with $1/j\omega C$ is

$$\frac{R \times 1/j\omega C}{R + 1/j\omega C} = \frac{R}{1 + j\omega CR} = \frac{R(1 - j\omega CR)}{1 + (\omega CR)^2}$$

$$= \frac{100[1 - j2\pi \times (50 \times 10^6) \times (20 \times 10^{-12}) \times 100]}{1 + [2\pi \times (50 \times 10^6) \times (20 \times 10^{-12}) \times 100]^2}$$

$$= \frac{100(1 - j0.628)}{1.395} = 71.7 - j45 \, \Omega$$

$$\rho = \frac{Z_{\mathrm{L}} - Z_0}{Z_{\mathrm{L}} + Z_0} = \frac{71.7 - j45 - 78}{71.7 - j45 + 78} = \frac{-6.3 - j45}{149.7 - j45}$$

$$|\rho| = \sqrt{\frac{6.3^2 + 45^2}{149.7^2 + 45^2}} = \mathbf{0.29}$$

2.9 The return loss at the receiver is

$$20 \log 0.29 = -10.8 \, \mathrm{dB}$$

The pulses of radiation lose $2 \times 16.4 = 32.8 \, \mathrm{dB}$ in either direction to and from the transmitter, so the return loss at the transmitter is

$$-10.8 \, \mathrm{dB} - 32.8 \, \mathrm{dB} - 32.8 \, \mathrm{dB} = -76.4 \, \mathrm{dB}$$

The amplitude of the reflected pulses at the transmitter relative to the transmitted pulses is

$$10^{-76.4/20} = \mathbf{1.5 \times 10^{-4}}$$

4.1 Assuming 854 MHz to be 90 per cent above f_c,

$$f_c = 854/1.9 \approx 450 \, \text{MHz}$$

For this value of f_c, $614/450 = 1.36$, so the lowest frequency is 36 per cent above f_c; well out of the high attenuation zone.

For $f_c = 450 \, \text{MHz}$, $\lambda_c = (3 \times 10^8)/(450 \times 10^6) \, \text{m} = 66.7 \, \text{cm}$. $\lambda_c = 2a$ (a being the wide guide dimension) so this indicates $a = 33.3 \, \text{cm}$. The value is not critical, so if we say $a = 32 \, \text{cm}$, this gives $\lambda_c = 64 \, \text{cm}$, $f_c = (3 \times 10^8)/0.64 = 468.75 \, \text{MHz}$, the lowest frequency in the band is $614/468.75 = 1.31 = 31$ per cent above cut-off and the highest frequency $854/468.75 = 1.82 = 82$ per cent above cut-off which is satisfactory. So an appropriate solution for the guide dimensions is

32 cm by 16 cm

4.2 $\lambda_c = 2a = 8 \, \text{cm}$

$$\lambda = c/f = (3 \times 10^8)/(6 \times 10^9) \, \text{m} = 5 \, \text{cm}$$

$$\frac{1}{\lambda_g^2} = \frac{1}{\lambda^2} - \frac{1}{\lambda_c^2} \tag{4.4}$$

so

$$\frac{1}{\lambda_g^2} = \frac{1}{25} - \frac{1}{64} = 0.024\,375$$

giving $\lambda_g = \mathbf{6.405 \, cm}$

$$v_{p(\text{guide})} = (\lambda_g/\lambda) \times c \tag{4.1}$$

$$= (6.405/5) \times (3 \times 10^8) \, \text{m/s}$$

$$= \mathbf{3.843 \times 10^8 \, m/s}$$

$$v_{g(\text{guide})} = (\lambda/\lambda g) \times c \tag{4.2}$$

$$= (5/6.405) \times (3 \times 10^8) \, \text{m/s}$$

$$= \mathbf{2.342 \times 10^8 \, m/s}$$

$$Z_W = 120\pi(\lambda_g/\lambda)$$

$$= 120\pi \times 6.405/5$$

$$= \mathbf{483 \, \Omega}$$

5.1 Using Equation (5.1)

$$d < \frac{2.405\lambda}{\pi(n_1^2 - n_2^2)^{1/2}}$$

$$d < \frac{2.405 \times 0.85}{\pi(1.460^2 - 1.458^2)^{1/2}}$$

giving $d < 8.5\,\mu m$. The choice is not critical, since we are nowhere near the dispersion minimum, so, for the sake of mechanical tolerances, choose say, **$d = 7.5\,\mu m$**.

5.2 X 2 km lengths would require $X - 1$ splices. Assume again 0.5 dB loss per splice (although the smaller core diameter makes the splices a bit harder to make accurately), 2 dB for terminal connectors and 5 dB repair margin. The system margin is 41 dB.

At the limit,

$$3X\,dB + 0.5(X - 1)\,dB + 2\,dB + 5\,dB = 41\,dB$$

gives

$X = 9.86$ lengths.

The choice would be 9 lengths (or 10 if you want to shade the repair margin a bit), so the maximum span is

18 km

Power budget:

transmitter power	−3 dBm
receiver sensitivity	−44 dBm
System margin	41 dB
fibre loss	27 dB
splice loss	4 dB
connector loss	2 dB
repair margin (10 splices)	5 dB
Route losses	38 dB
Excess margin	3 dB

5.3 For a signalling rate of 36 Mbaud, the symbol period is

$$\frac{1}{36 \times 10^6} = 27.8\,ns$$

so the dispersion must be kept below 13.9 ns.

For a span of 18 km, the dispersion is

$60 \times 18 \times lw$ ps

where lw is the linewidth of the source in nm

$1080 \, lw < 13.9 \times 10^3$

lw < 12.8 nm

indicating that a laser should be used.

5.4 Using the formula for intermodal dispersion

$$\frac{Ln_1}{cn_2}(n_1 - n_2) < 13.9 \times 10^{-9}$$

$$L < \frac{(13.9 \times 10^{-9}) \times (3 \times 10^8) \times 1.458}{1.460(1.460 - 1.458)} < 2082 \, m$$

The maximum span is **2 km**.

No splices would be needed initially, so allowing 2 dB for couplings (probably less, since coupling is easier with a wide core) and say 3 dB for repairs, plus the attenuation of 2×1.5 dB, the total route losses would only be 8 dB. With a receiver sensitivity of the order of -40 dBm or better, a source power of -13 dBm would be more than adequate.

6.1 Link budget ship to land station:

Ship to satellite	
EIRP	36 dBW
At. abs.	-0.2 dB
Free-space p.l.	-188.9 dB
$1/k$	228.6 dB
G/T	-13.2 dBK
$(C/N_0)_u$	62.3 dBHz

Satellite to land station	
EIRP	-2.5 dBW
At. abs.	-0.4 dB
Free-space p.l.	-197.2 dB
$1/k$	228.6 dB
G/T	32.0 dBK
$(C/N_0)_d$	60.5 dBHz

Channel $N_0/C = (N_0/C)_u + (N_0/C)_d$

-62.3 dB $= 5.89 \times 10^{-7}$

-60.5 dB $= 8.91 \times 10^{-7}$

$\quad$ sum $= 1.48 \times 10^{-6} = -58.3$ dB

Channel $C/N_0 = $ **58.3 dBHz**

The satellite EIRP is kept as small as possible to minimize satellite power consumption: even with this value of EIRP the ship to shore link C/N_0 ratio is better than the shore to ship by about 4 dB.

6.2 The main beamwidth, between half-power points, in radians, is approximately equal to the inverse of the diameter of the effective aperture in wavelengths.

At 6.42 GHz, $\lambda = (3 \times 10^8)/(6.42 \times 10^9) = 0.047$ m.

The effective aperture diameter may be taken to be 80 per cent of the physical diameter of the paraboloid, i.e. $0.8 \times 13 = 10.4$ m so

beamwidth $\approx 0.047/10.4$ rad $= 4.5 \times 10^{-3}$ rad

$$= (4.5 \times 10^{-3}) \times 180/\pi^\circ = \mathbf{0.26^\circ}$$

At 4.2 GHz, $(3 \times 10^8)/(4.2 \times 10^9) = 0.071$ m wavelength,

$0.071/10.4 = 6.87 \times 10^{-3}$ rad $= \mathbf{0.39^\circ}$

The beamwidth limits represent 3 dB reduction on either side of the centre of the beam. Thus the smaller of the two beamwidth values, 0.26°, suggests a need for an antenna pointing accuracy of the order of $\pm \mathbf{0.1^\circ}$.

6.3 At the land station, EIRP/gain = transmitter power.

$$G = (4\pi A')/\lambda^2$$

Taking A' as 2/3 the physical aperture of the antenna

$$G = \frac{4\pi \times (2/3)\pi(6.5)^2}{[(3 \times 10^8)/(6.42 \times 10^9)]^2} = 5.5 \times 10^5 = 57 \text{ dB}$$

so transmitter power $= 60$ dBW $- 57$ dB $= 3$ dBW $= \mathbf{2\,W}$.

There is no point in transmitting more power since the uplink carrier to noise-density ratio is already about 15 dB better than that of the downlink.

6.4 $G = (4\pi A')/\lambda^2$.

Taking A' as 2/3 the physical aperture of the antenna, at 1.64 GHz

$$G = \frac{4\pi \times (2/3)\pi(0.6)^2}{[(3 \times 10^8)/(1.64 \times 10^9)]^2} = 283 = 24.5 \text{ dB}$$

Ship transmitter power = EIRP/gain $= 36$ dBW $- 24.5$ dB

$$= 11.5 \text{ dBW} = \mathbf{14\,W}$$

At 1.54 GHz, ship's antenna gain is

$$\frac{4\pi \times (2/3)\pi(0.6)^2}{[(3 \times 10^8)/(1.54 \times 10^9)]^2} = 250 = 24 \text{ dB}$$

Since $G/T = -3.5$ dB, $T/G = 3.5$ dB so

$(T/G) \times G = 3.5$ dB $+ 24$ dB $= 27.5$ dBK $= 562$ K

The equivalent noise temperature is **562 K**.

Since the antenna is looking up at the sky, most of this noise is produced in the receiver.

For solutions 7.1 to 7.5 see Fig. S1.

7.1 The normalized impedance represented by point A is

$0.3 + j0.7$

so the impedance of the load is this multiplied by Z_0, i.e.

$30 + j70\,\Omega$

7.2 The normalized admittance of the load is represented by the point B, and so is approximately

$0.5 - j1.2$

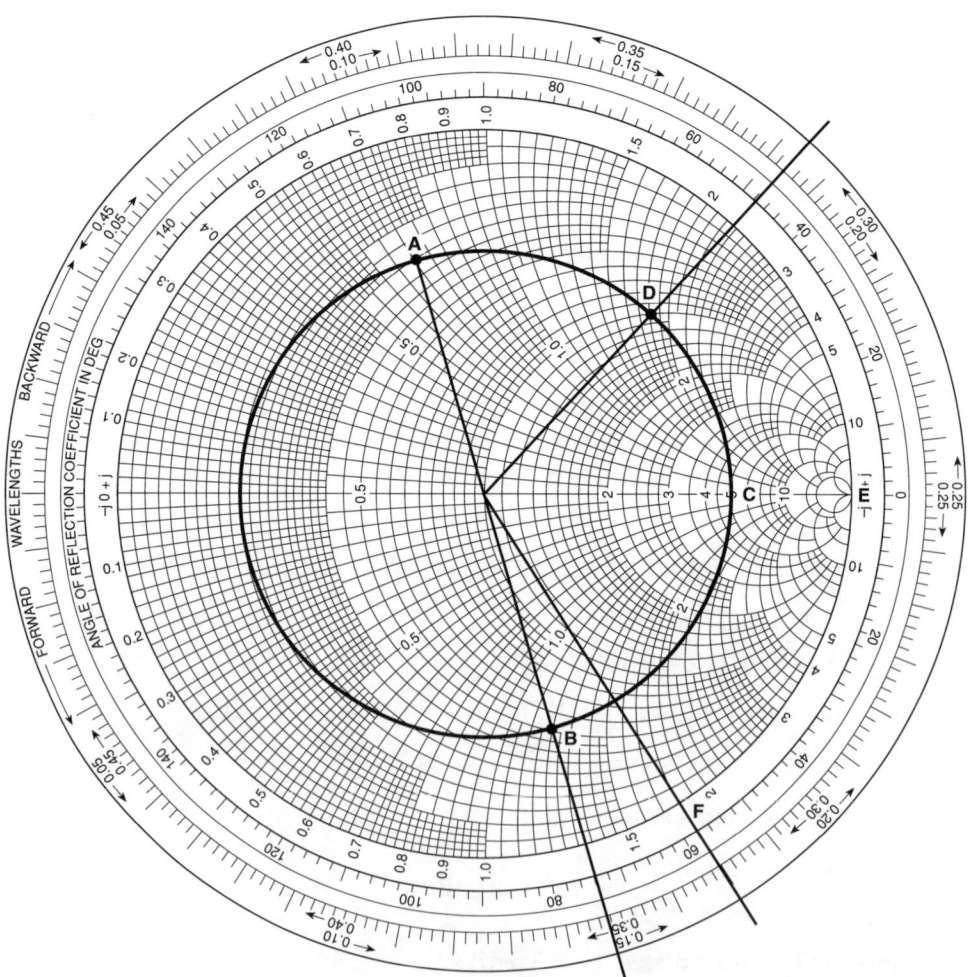

Fig. S1

7.3 The VSWR can be read off at point C as

5 : 1

7.4 Moving backwards from the load admittance at B, the first minimum is at C. Measuring round the circumference of the chart gives

0.398λ

so the distance from the load to the first minimum is

$0.398 \times 15 \approx$ **6 m**

7.5 The nearest point to the load for a stub is represented by D. The distance round the chart from B to D is

$0.332\lambda = 0.332 \times 15 \approx 5\,\mathrm{m}$

The normalized susceptance at D is j1.8.

The required stub length to give a normalized susceptance of $-$j1.8 is represented by the arc E-F, which represents

$0.08\lambda = 0.08 \times 15 \approx 1.2\,\mathrm{m}$

A shorted stub **1.2 m** long must be placed **5 m** from the load.

Index